The Yorkshire Dales

LANDSCAPE AND GEOLOGY

The Yorkshire Dales

LANDSCAPE AND GEOLOGY

Tony Waltham

THE CROWOOD PRESS

First published in 2007 by
The Crowood Press Ltd
Ramsbury, Marlborough
Wiltshire SN8 2HR

www.crowood.com

British Library Cataloguing-in-Publication Data
A catalogue record for this book is available from the British Library.

ISBN 978 1 86126 972 0

Acknowledgements
Some of the maps and cave surveys were made from originals by the author, but most have been abstracted, compiled and redrawn from a variety of sources. Due credit belongs to members of the British Geological Survey for most of the geological information, and to a host of local cavers for most of the cave surveys.

The author is grateful to friends who supplied photographs – Tony Baker (130), Rob White (161 and 172), John Dale (167 bottom and 170 bottom), the late Oliver Statham (126 top) and Hanson PLC (176). Harry Long, Dave Lowe and Jerry Wooldridge are thanked for extensive help while preparing the book.

The book is dedicated to the author's wife Jan, who acted as scale for so many photographs, and whose boundless support made it all possible.

Series editor: Peter Doyle

Frontispiece: The stone houses of Kettlewell nestle on the floor of Wharfedale beneath limestone scars that rise toward the ridge of Great Whernside.

Typeset by Jean Cussons Typesetting, Diss, Norfolk

Printed and bound in India by Replika Press

Contents

Introduction

Limestone scars, stone-built villages, open grassland, deep wide valleys, sheep farms, moors of grit. Any of these could be claimed as characteristic, but in fact they all combine to create the beautiful and distinctive landscapes of the Yorkshire Dales. Created by the rocks, moulded by erosion and trimmed by man, this slice of the northern Pennines offers some of the finest countryside in England.

It is this special blend of nature and man that makes the Yorkshire Dales such a richly varied landscape, where each valley, each turn, each hill reveals a new aspect of a truly splendid terrain. Within the Dales, there are three clear sections, each defined by its own rock type. In the south, the Craven Dales include the limestone plateaux around the Three Peaks, Malham and Wharfedale. The northern Dales make the ultimate contrast with their

Pen-y-ghent stands clear in the long view across Ribblesdale from the limestone pavements of Ingleborough.

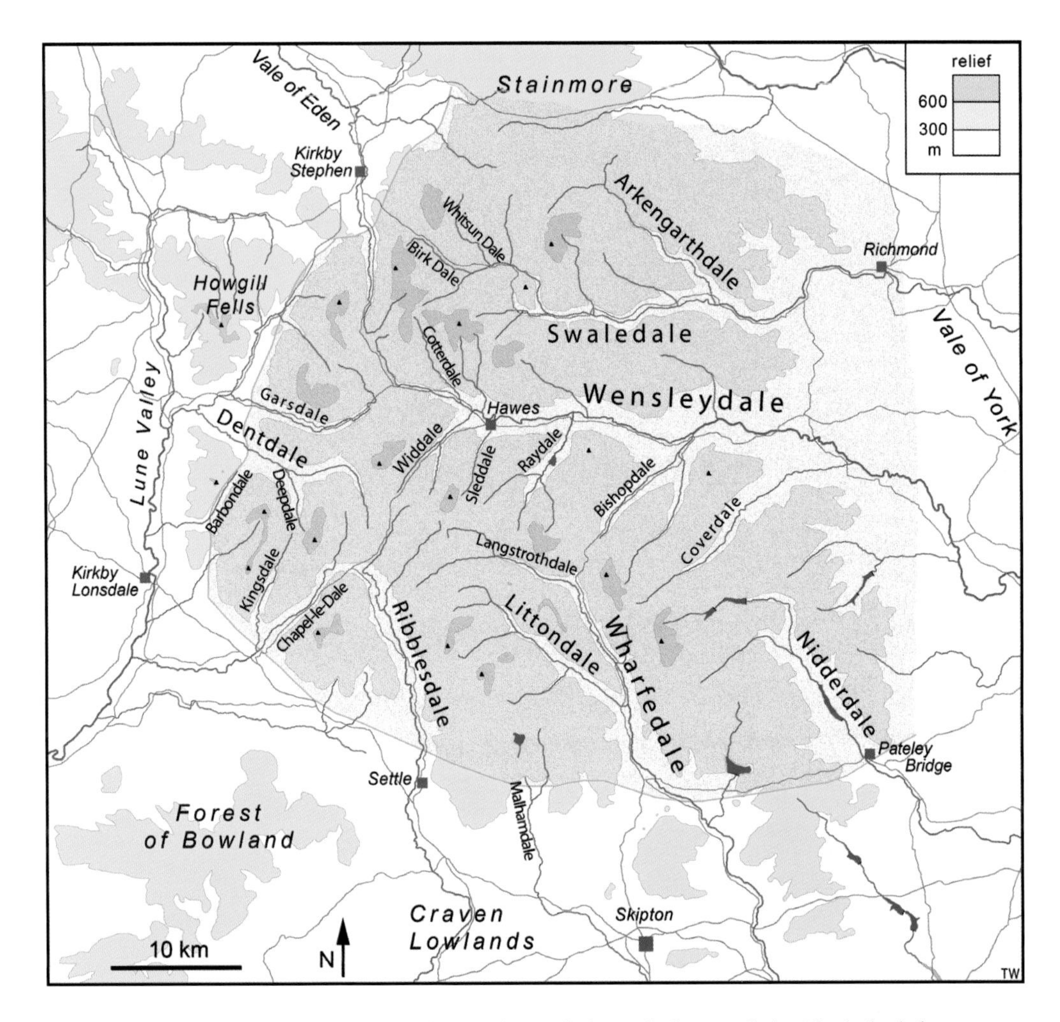

The main landscape features of the Yorkshire Dales, with the underlying Askrigg Block shaded green.

grit moors either side of Swaledale. And in between, the benched hillsides of Wensleydale are cut into a mixture of rocks known as the Yoredales. Plate movements and seabed deposition 300 million years ago created these rock types and defined the Dales of today.

Superimposed on those rock landscapes are the results of a million years of erosion, which climaxed in the Ice Ages. The Dales themselves are individual glaciated troughs that were filled with ice until about 15,000 years ago; they are just the most conspicuous results of the Ice Age glaciers whose imprint is felt across almost the entire Dales landscape. Time scales become shorter, but within the last few thousand years man has had a huge impact on the details of the Dales landscapes. Man has created the third element – the texture that is draped over the fells and dales, which had already been scoured from the great sequence of solid rock. Walls and houses, mines and quarries may not be natural, but they are inextricable components of the landscapes. Man's interface with geology is rarely seen as clearly as in the Dales.

Any story of the Yorkshire Dales landscapes inevitably makes frequent reference to the Askrigg Block – which had, quite literally, a

ABOVE: Dale Head Farm has limestone pavements in front and the whaleback of Pen-y-ghent behind.

BELOW: Swaledale is a sea of green and yellow in the early summer.

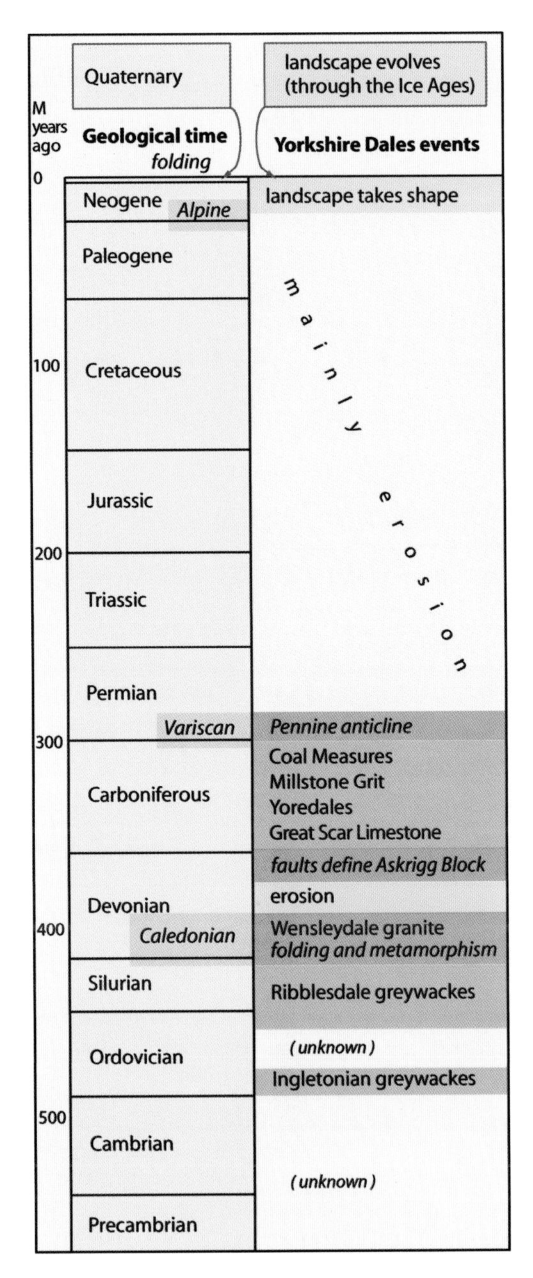

The Yorkshire Dales through geological time. Most of the recognizable landscape features were developed during the last million years, within the Quaternary, long after almost all the rocks and rock structures had been created between 500–300 million years ago. More detail of the Quaternary events is given in the table on page 68, and Carboniferous events are shown in the maps on page 59.

Stone houses line the narrow village street in Langthwaite, Arkengarthdale.

very deep background influence. The entire Yorkshire Dales area sits on a chunk of ancient 'basement' rock that is known as the Askrigg Block. Though little of this is seen at the surface, it dictated the way in which the Carboniferous rocks were laid down as sediments on ancient sea floors – much later to be carved into today's distinctive landscapes. The Askrigg Block still defines the boundaries of the Dales region. Bounded by faults, deep fractures in the rock, its marginal Craven Faults border lowlands to the south, while its Dent Fault borders the Howgill Fells and

Cliffs of Great Scar Limestone fringe Chapel-le-Dale, with Ingleborough's dark summit beyond, and an old quarry exposing the basement rocks of the Askrigg Block below.

Cascading water carved most of the Dales landscape – here tumbling over jointed limestone at Aysgarth Falls in Wensleydale.

Lake District to the west. The ice-scoured Stainmore Gap is the break from the grit expanses of the northern Pennines, and grit moors to the east descend steadily off the Block towards the Vale of York.

Development of a landscape is a long, complex and fascinating process. It evolves through a sequence of rocks and continues into the processes of erosion. Different elements interact, and that is why each landscape has so much individual character; few have more of that than the Yorkshire Dales. Consequently, readers will find references to other sections as they read through these pages. There is a purpose to these, as the story of the Dales is a bit like knitting – only by looping back every so often does it all hang together. Not everything is yet known about the Dales landscape. Just how and just when some features were created retains an air of mystery. Even the exact origins of the two landforms perhaps best known – Malham Cove and Gaping Gill – are still hotly debated. These pages can only hope to report current thinking, and raise just some of the questions that deserve further investigation. And perhaps those can provide yet more reason to keep returning to the glorious Dales.

The magnificent glaciated trough of Wharfedale, just up from Kettlewell.

Part I: Starting with the Rocks

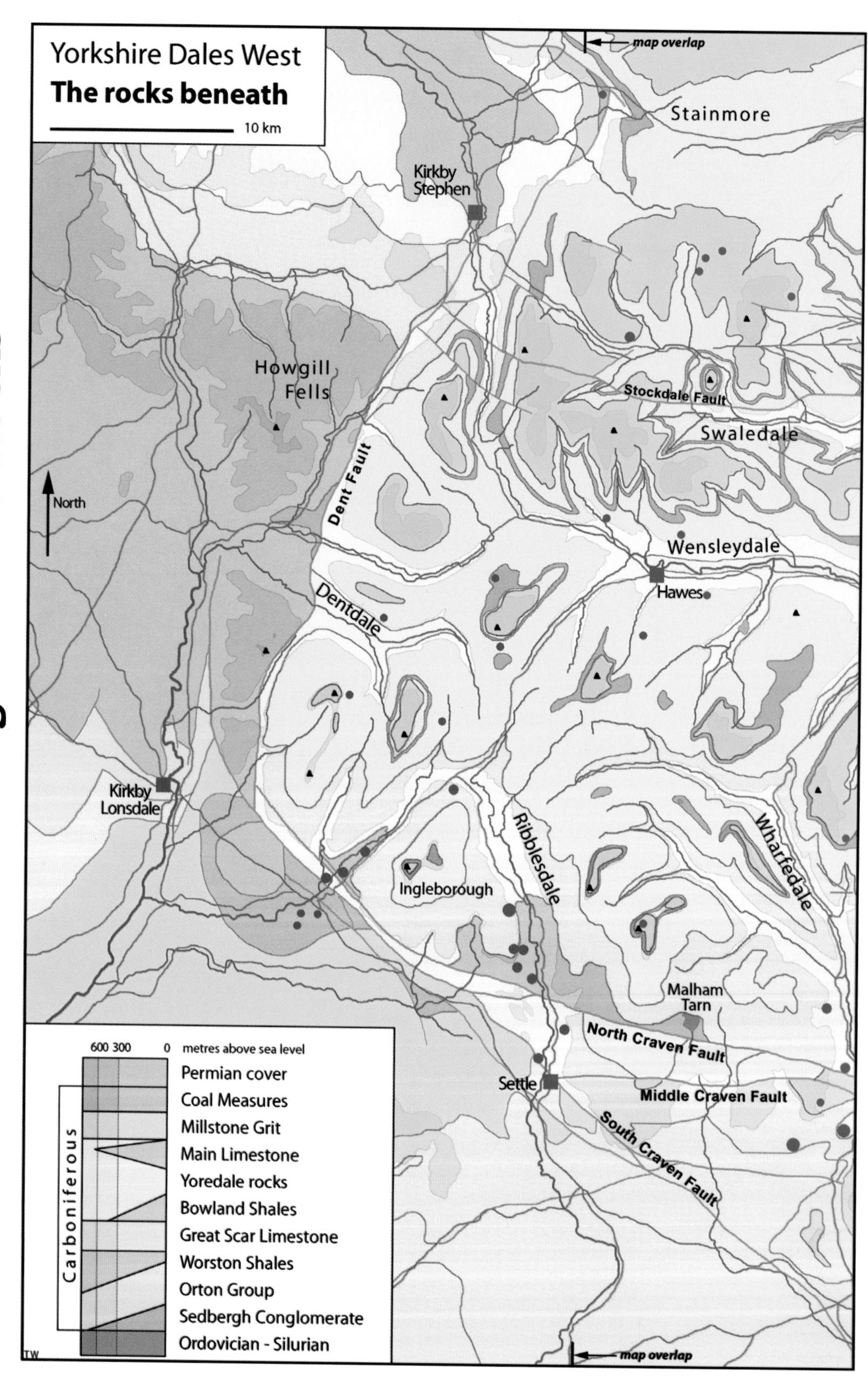

Part I: Starting with the Rocks

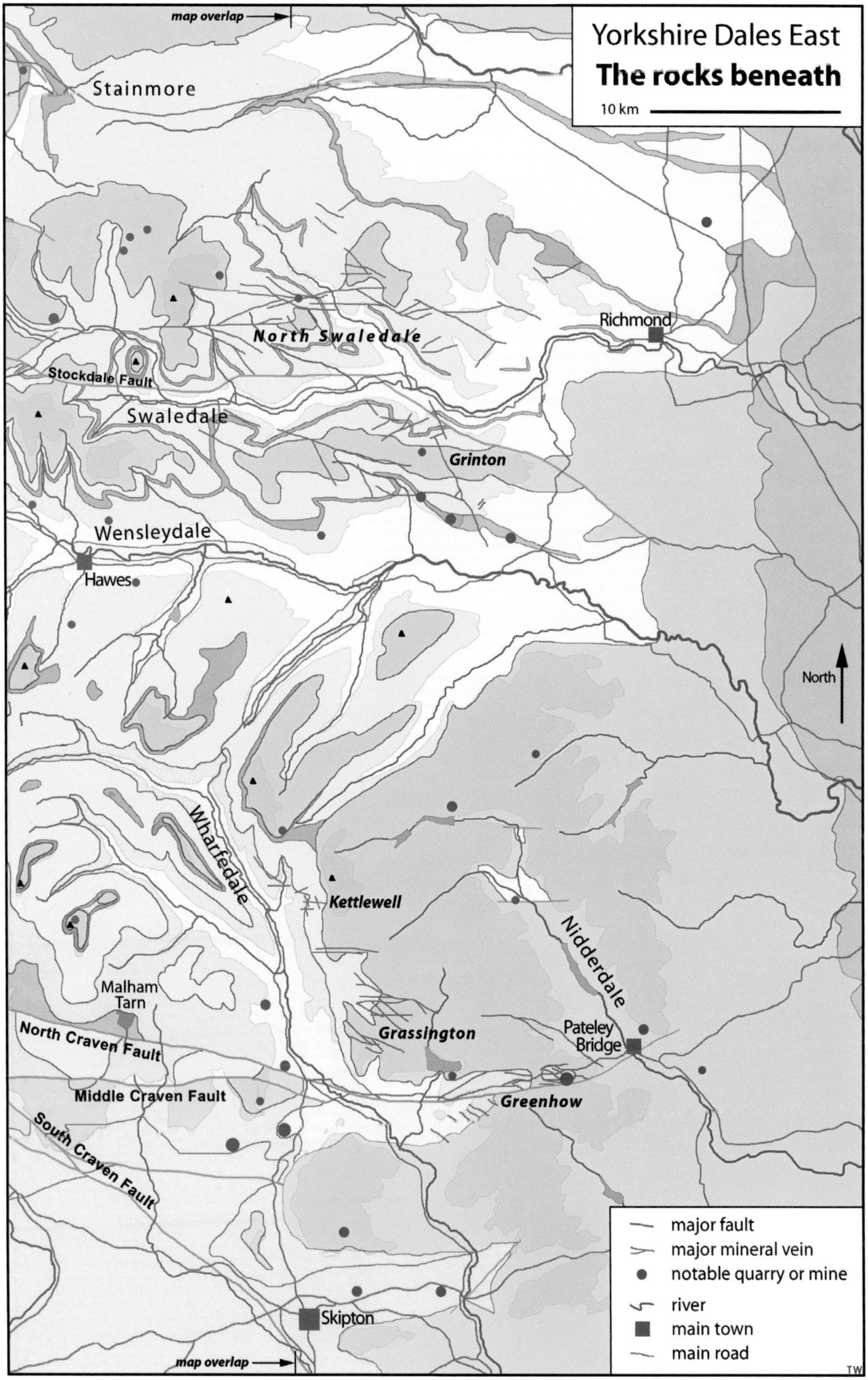

Yorkshire Dales East
The rocks beneath
10 km

map overlap
Stainmore
North Swaledale
Richmond
Stockdale Fault
Swaledale
Grinton
Wensleydale
Hawes
North
Wharfedale
Kettlewell
Nidderdale
Malham Tarn
Grassington
Pateley Bridge
North Craven Fault
Middle Craven Fault
Greenhow
South Craven Fault
Skipton
map overlap

major fault
major mineral vein
notable quarry or mine
river
main town
main road

TW

CHAPTER 1

Deep Beginnings

Basement is a general term for the oldest known rocks that underlie a region. Most basement rocks are partly or completely hidden beneath the cover of younger rocks, but they are important because their structures dictate so many of the structures in the rocks above them, and therefore influence many of the features in today's landscapes.

Within the Yorkshire Dales, just a slice of the basement is visible along the southern edge of the Askrigg Block – largely where short sections of Ribblesdale and Chapel-le-Dale have been cut down into it. Excavation of the Dales created valley-floor features called inliers that are ancient rocks surrounded by outcrops of the younger rock; these inliers are windows into the deeper geological structure. They are windows into the basement that extends beneath the entire region, but which lies unseen beneath covering rocks that remain elsewhere. In a contrast of depths, the top surface of the basement rocks (and therefore the base of the Carboniferous rocks), is exposed at the surface in Ribblesdale, but lies buried 2000–5000m (6500–16,400ft) beneath ground level in the Stainmore Trough and the Craven Basin, respectively just north and south of the Askrigg Block.

Ingletonian basement

The oldest rocks now visible in the Dales are basement beds that lie within the Ingleton

Silurian		**Neals Ing Formation**	greywackes
	Ludlow	**Studfold Sandstone**	greywackes
		Horton Formation	flagstones, siltstones
	Wenlock	**Austwick Formation**	greywackes
Ordovician	*Ashgill*	**Coniston Group**	mudstones, sandstones
	Arenig	**Ingleton Group**	greywackes, slates

The main units of Ordovician and Silurian rocks exposed in the inliers of Ribblesdale and along the southern edge of the Yorkshire Dales.

Ingletonian greywackes exposed on the working benches of the Ingleton Quarry. The tight syncline separates vertical beds to the south from steeply dipping beds to the north, both of which form some of the waterfalls in the adjacent Ingleton glens, though the fold is difficult to see in the stream sections.

Group. They have no visible fossils, so for many years they were thought to be of Precambrian age. But they have now been dated by an analysis of their decaying radioactive isotopes, and are known to about 480 million years old, which places them in the early Ordovician age. They form the valley-floor inliers in Chapel-le-Dale and lower Kingsdale, and also a small outcrop around Horton in Ribblesdale, and their slates and greywackes are so strongly folded that most of them are nearly vertical at outcrop.

Greywacke is a type of strong, coarse-grained sandstone, and its beds form the stepped cascades of Pecca Falls, where the River Twiss descends from Kingsdale towards the Craven Faults (*see* page 54). In Chapel-le-Dale, the same beds have been extracted for roadstone (*see* page 187), where a viewing platform provides a splendid view down the terraced benches of the deep Ingleton Quarry.

Both valleys contain a number of small river-bank quarries within the Ingleton Glens, where the slates were worked for roofing material in the past. All these slate quarries are found in a single band of the best material, which is repeated in the eastern glen because it is wrapped around the steep synclinal fold of greywacke that is quarried just east of the wooded glen. Continuing up Chapel-le-Dale, the Ingletonian rocks appear to have almost no repetition by folding and are therefore about 3000m (10,000ft) thick. This includes a narrow band of coarser greywacke in which gravel-sized chunks of pink feldspar led to it being mistakenly called granite, when it was exposed in the old 'Granite Quarry' midway up the dale.

These coarse grained Ingletonian rocks are known as greywackes because they were formed in a distinctive deep-water environment, which is recognized by some inherited

Once an Ocean Floor

The oldest rocks seen in the Dales originated as sediments that were laid down near the South Pole. Around 500 million years ago, much of our planet's land area was all in one piece, on the supercontinent known as Gondwana. But giant Earth movements – plate tectonics – were active even then, and a chunk of continental crust known as Avalonia broke away. It now forms the deep basement of part of Central Europe, including the Midland Platform of old rocks that underlie central England. Sand and mud carried by rivers that drained off Avalonia accumulated on the floor of the ocean between it and Laurentia (the other supercontinent of the time, which now underlies most of North America). Buried beneath further sediments, these sands and muds were turned into sandstones and shales that were eventually to become the Ingletonian rocks within the Dales basement.

That ocean (where the oldest known rocks of the Yorkshire Dales were formed) was not the Atlantic, but the Iapetus Ocean. Iapetus was named by its discoverers after the father of Atlas, the king of Atlantis in Greek mythology. This far back in time, the Atlantic Ocean was not even a twinkle in a geological eye. Before the Carboniferous, the section of the world's crust that later included Britain was dominated by the Iapetus Ocean, with a rapidly subsiding floor where sediments kilometres thick accumulated along an active convergent tectonic plate boundary. Sand and mud were carried into these deep waters in series of submarine landslides that created dense, sediment-rich turbidity flows. From each successive turbidity cloud, the sand settled out first, followed by the finer mud, to create sequences where each sandstone grades up into a finer mudstone. These rocks are known as turbidites, with each couplet of coarse and fine beds perhaps a metre thick.

The beds of the Ingleton Group are all turbidites. They were caught in the plate convergence in late Ordovician times, and were crumpled into isoclinal folds (with beds on both of the sides almost parallel to each other) and were lightly metamorphosed. The increase in heat and pressure turned the sandstones into tough greywackes and the mudstones into poor quality slates. The original proportions of sand and mud varied, so that slate-rich beds have been worked for roofing slates, while a greywacke-rich zone is extracted for roadstone at the Ingleton quarry.

In Silurian times, more turbidites were deposited in a shelf sea across the top of the buried Ingletonian, and these too were then caught up in the plate convergence. The plate movements were a continuous process, and finally led to the complete closure and disappearance of the Iapetus Ocean (and more than a hundred million years passed before the Atlantic Ocean started to open obliquely across the line of its ancestor's demise). This was the last time that the Yorkshire Dales region was caught up in the heart of major orogenesis – a belt of mountain building (orogenesis means mountain beginning in Greek) where strong folding and metamorphism are among the many mountain building processes that develop where plates collide, and which are so important in keeping the geology alive and renewable in a living planet.

A single unit of turbidite about a metre thick and dipping steeply to the right, exposed on Newfield Crags near Helwith Bridge. The unit has a thick bed of coarse greywacke that grades up (to the right) into finer-grained slate. Both bounding bedding planes (each in shadow) have flute casts where the greywacke base is indented into the slate top.

Spectacular flute casts in the Silurian turbidites on Newfield Crags near Helwith Bridge. This is the underside of an original bed of sand that filled the flute hollows scoured into the underlying clay by a turbidity flow across the sea floor. Now folded into a vertical position, the casts in the sandstone have been exposed by erosion and removal of the weaker slate.

The Silurian turbidites are comparable to those in the Ingleton Group, and some do show splendid flute casts. These are elongate hollows, perhaps 300mm (12in) long, that were scooped out of the soft clay sediment by turbidity flows that swept over them, and were then filled by sand settling out of the turbidity cloud. They are now preserved as casts, which appear as gentle bulges on the underside of greywacke beds, where the slate has been removed by weathering; so they are best seen in exposures of nearly vertical beds, and those on Newfield Crags near Helwith Bridge are perhaps the finest to be seen. They are the ultimate evidence of the sedimentary environment in that by-gone ocean. It was a time of major geological activity, when many of the individual turbidity flows were probably triggered by earthquakes. Submarine banks of unstable sediment were derived from land and accumulated on shelves near to shore and in shallow water; with an earthquake, they collapsed, slumped and slipped into deeper water, breaking into turbidity clouds almost like an avalanche of powder snow, except that these were far larger and flowed deep underwater.

With outcrops only in Ribblesdale, the Horton Flags have a layering derived from gentle pulses in the deposition, but the powerful turbidity flows returned later to create the coarser Studfold Sandstones (a bed within the Horton Flags) and then the Neals Ing Formation, the youngest of the Silurian rocks in the Dales inliers. The Iapetus Ocean then finally closed, and the sediments on its floor were scraped up by the over-riding continental plate, and crumpled and metamorphosed into a folded rock sequence. With half of them lost to erosion in the Devonian, those that remained became the floor of the Carboniferous sea and the basement of today's Askrigg Block.

structures within them (*see* page 17). The slates are a finer-grained equivalent of the coarse sediments formed in the same environment, and both rocks have been metamorphosed to a low grade, and that is why they are so hard and strong.

Ordovician Ingletonian rocks underlie an unknown portion of the Askrigg Block, but they have been proven in just one borehole, at Beckermonds Scar in Upper Wharfedale. This borehole was drilled in 1976 to check a strong magnetic anomaly that had been revealed by airborne geophysical surveys. It found a greywacke containing about 3 per cent of the mineral magnetite – the variety of iron oxide that is naturally magnetic. Though magnetite can be a valuable iron ore, at this site it is disseminated in such large volumes of rock that it could never be worth mining.

Silurian basement

Basement exposed in the Ribblesdale inlier is dominated by a sequence of Silurian rocks nearly 2000m (6000ft) thick, which were deposited within that period about 450–410 million years ago. The lowest Silurian unit is the Austwick Formation, which is dominated by greywackes similar to those in the Ingletonian. This has an outcrop from Crummack Dale to south of Helwith Bridge and beyond. In a wedge between these and the North Craven Fault, and lying between Austwick and Wharfe, are Upper Ordovician sandstones, siltstones and mudstones, some of which originated as debris from explosive volcanoes. Both these Silurian and Ordovician beds contain scattered layers that contain graptolites and fragments of trilobites, but the fossils are difficult to find, and many of them are poorly preserved. The Austwick Formation greywackes form the ribs of strong rock on the west flank of Crummack Dale that were the sources for the well-known glacial erratics that lie on Norber Scar (*see* page 88).

The thickest of the Silurian beds is the Horton Formation, which consists mainly of well-laminated siltstones. These have been quarried in Ribblesdale, first for the flags that are so visible in the walls and buildings around Helwith Bridge, and latterly for aggregate in the same area (*see* page 188). The same siltstones also form the watertight floor of Malham Tarn, though there they are almost totally obscured by drift deposits and the Tarn itself. All these rocks now lie in steep folds, with a central syncline and adjacent anticlines, all aligned almost parallel to the North Craven Fault. Siltstones very similar to those of the Horton Formation were also found in the Chapel House borehole in Wharfedale.

Wensleydale Granite

Beyond the inliers along the southern edge, and a few boreholes, the nature of the Askrigg Block basement remains unseen and unknown. However, geologists long suspected that a granite lay at depth. Because granite is a rock of relatively low density, this would account for the Block 'floating' up to its high position in the regional structure, and molten granite could also have provided a source for the lead minerals that are so rich in the Swaledale area.

So, in 1973, a deep borehole was drilled in Raydale – and the Wensleydale Granite was discovered nearly 500m (1600ft) below ground level. It is a pink, medium-grained rock, rather similar to some of the granites exposed in the Lake District. But analysis of its unstable radioactive isotopes showed that it is about 400 million years old, dating from the Devonian period, and much earlier than both the Carboniferous limestones and their mineral veins. So it was not the source of the minerals, but it did account for the Askrigg Block being a relatively high area when it was a shallow-water platform beneath the Carboniferous seas. That granite completes our picture of the deep basement, and established the structures that dictated the formation of the next generation of sedimentary rocks – the limestones of the Carboniferous.

CHAPTER 2

White Limestone

Strong white rock creates landscapes of the finest quality. Towering cliffs, bright enough to enhance the landscape instead of enclosing it, provide drama in the hills that stand between the dales. Add to this the special features of gorges, pavements and caves that come in a mature karst landscape (*see* page 102) and limestone even outclasses granite, its main contender in the mountain beauty competition.

The limestones of the Yorkshire Dales are of Carboniferous age, which means that they are between 360–300 million years old. The Carboniferous name refers to the coal found in the upper part of the great sequence of rocks of that age, and the Dales' limestones are all to be found in its lower part. The name does not imply that there is coal or free carbon in the limestone; though of course there is carbon locked into the calcium carbonate – the mineral calcite, which makes up all the limestone except for the minor clay impurities.

Great Scar Limestone

Landscapes in the southern half of the Yorkshire Dales are dominated by outcrops of

The white crags and scree slopes that form Twisleton Scars in the Great Scar Limestone along the side of Chapel-le-Dale.

Great Scar Limestone. The cliffs of Malham, the benches of Ingleborough and the scars of Wharfedale are all made of this one strong, white, Carboniferous Limestone.

Targeted by quarrymen because it is so strong, and beloved by cavers because it provides the home for their sport, the limestone is hardly welcomed by fossil hunters because it is surprisingly featureless and mostly devoid of collectible fossils. Though made almost entirely of broken seashells, nearly all the rock is now a fine-grained mass of calcite crystals, whose growth and intergrowth destroyed most recognizable fossils when it was lithified, from a soft seabed mud into a solid limestone. Furthermore, most of the Great Scar is actually pale creamy grey or even dark grey in colour, its white appearance in the landscape due only to a thin patina created by weathering.

The Great Scar Limestone reaches to more than 200m (650ft) thick at outcrop around the southern Dales, and nearly the full sequence is exposed along the cliffs of Chapel-le-Dale and around Gordale Scar. All these cliffs are scored by almost horizontal bedding planes, because the limestone is all in beds 1–10m (3–33ft) thick. Each major bedding plane is actually a paper-thin layer of shale, which is easily removed by weathering, so that it is rarely seen except underground, where it can usually be picked out in the clean-washed walls of stream caves and shafts. Each bed of limestone is also broken by joints – natural breaks in the rock that were created when the whole rock mass has all been gently deformed in Earth movements. Most joints are close to vertical, and they dictate the precipitous profiles of the cliff faces, of which Malham Cove is just the highest of many. Intersecting sets of joints create the cross-hatch patterns on the big limestone pavements, though an excess of joints leaves the rock too broken, easily weathered and soon lost beneath soil.

Basement rocks were folded during the Caledonian Orogeny, then uplifted and eroded down to a rolling plane. This is now seen as the basal unconformity of the limestone, a time gap that is famously exposed at Thornton Force (*see* page 78) and clearly seen in the cliffs

Main units of the Great Scar Limestone in the Yorkshire Dales, and associated rocks in the basins north and south of the Askrigg Block. Like any area of sea bed over 50km (30 miles) across, the Carboniferous sea floors, in what is now the Dales area, varied in their depth and nature; so the rocks of today show the resultant lateral variations, and these tables of the rock sequences are greatly simplified.

Asbian	Worston Shales	Gordale Limestone	**Great Scar Limestone**	Danny Bridge Limestone Garsdale Limestone	Orton Group
Holkerian	Worston Shales	Cove Limestone Kilnsey Limestone	**Great Scar Limestone**	Fawes Wood Limestone	Orton Group
Arundian	Worston Shales	Basal conglomerate		Ashfell Sandstone Tom Croft Limestone	Orton Group
Chadian	Thornton Shale	***Askrigg Block***			
Courceyan	Chatburn Limestone	***Askrigg Block***		Sedbergh conglomerate	
	Craven Basin	***Askrigg Block***		***Stainmore Gulf***	

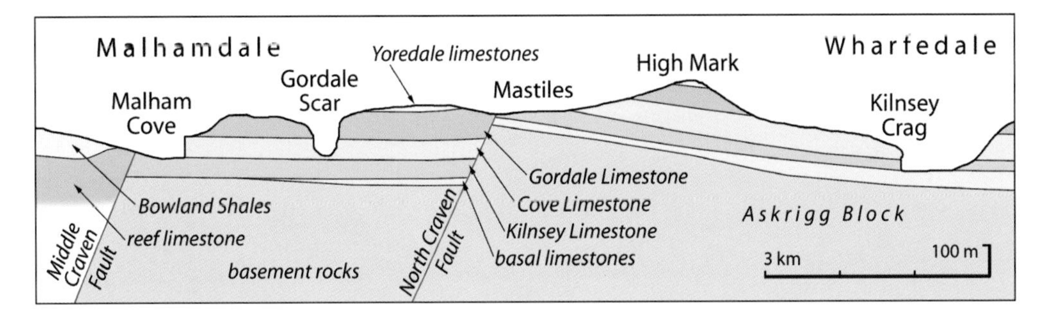

Profile from Malhamdale to Wharfedale (not drawn along a straight line) to show the three main units of the Great Scar Limestone at their type localities. Vertical scale is four times the horizontal scale, so the dips on the Askrigg Block are less than they appear here.

of Moughton Scar above the road from Austwick into Ribblesdale. Although there is some basal conglomerate, a leftover from the time when the basement was being eroded, the lowest of the limestone beds are the Chapel House Limestone (which is seen in only a few small outcrops) and the Kilnsey Limestone, which forms the eponymous overhanging crag in Wharfedale. The rather grey Kilsney Limestone formed only on the low areas of the submerging Askrigg Block, so it is impure with mud washed from the higher parts that were still land at the time. On the northern part of the Block, a thick sequence of earlier Carboniferous rocks, only some of which are limestones, was formed in the Stainmore Gulf while the Askrigg Block was still dry land. These beds now lie deep beneath the Swaledale area, but they are exposed west of the Dent Fault, where limestones are interbedded with the Ashfell Sandstone in the Ravenstonedale area south of Kirkby Stephen.

Above the grey Kilnsey Limestone, the paler Cove and Gordale limestones form the best of the white cliffs that characterize the southern Dales. The Cove Limestone is the more massively (i.e. thicker) bedded; it forms the entire face of Malham Cove, except for the top few metres of Gordale Limestone, and forms the unbroken lower cliffs at Gordale Scar, beneath the rather more stepped cliffs of the Gordale Limestone above. At both sites, their boundary is a prominent bedding plane – it forms the main overhang high on Malham Cove. Around Ingleborough, the boundary is placed at the Porcellanous Band, a bed of very fine grained, structureless limestone, about 700mm (2ft) thick, that was probably formed in a very shallow lagoon where high evaporation accelerated calcite deposition. There are various thin beds of this very fine-grained limestone. They are difficult to recognize in surface outcrops, but one good exposure is where the Porcellanous Band appears as a pale-coloured bed not far above head height round the walls of the Main Chamber in the Gaping Gill Cave System (*see* page 140).

The top of the Great Scar Limestone is generally taken as the Girvanella Band. This is a very distinctive bed of dark platey limestone rich with pale, almond-sized nodules of a fossil alga called *Osagia* (but previously known as *Girvanella*), though it is quite difficult to find at outcrop. Locally the Girvanella Band splits into two adjacent bands, and its level has now been found to lie within the Hawes Limestone, the first of the Yoredale limestones, which, south of Wensleydale, sits without a break on the Gordale Limestone.

From the high cliffs of Malham, a gentle northward dip takes the Great Scar Limestone to ever-greater depths, so that it is only seen in parts of the floor of Wensleydale, and is not exposed at all in Swaledale. The Great Scar

Once a Tropical Sea

In early Carboniferous times, around 350 million years ago, Britain's piece of the Earth's crust lay just south of the Equator. Plate movements were ongoing, and extension of this part of the crust had broken the southern ramparts of the Caledonian mountains into fault-bounded blocks, the subsidence of which, to different levels, had created a complex pattern of seas and islands. The limestones that now dominate the southern Dales were formed as the coastal region steadily evolved – and the results have to be viewed in terms of events at successive intervals of time (each of which is given its own name by geologists, as in the table on page 20).

At first (in Courceyan and Chadian times), the southern Askrigg Block was dry land, but thick sediments were accumulating on its northern sector that was subsiding into the Stainmore Trough. These included limestones, but also thick beds of sandstone formed by material derived from erosion of the higher areas that were still above sea level. At the same time, some limestone and a mineral known as anhydrite (only known from boreholes near Malham, and formed by evaporation) formed in lagoons when the southern edge of the Block was intermittently submerged. The Askrigg Block overlooked the deep Craven Basin that was still subsiding just to the south. There the Worston Shales accumulated in deep water, along with their interbedded limestones that are now seen in the old quarries along the Haw Park ridge east of Skipton.

The Askrigg Block then slowly subsided (in Arundian times), until it was submerged. The

The colonial framework coral *Syringopora*, in a block about 50mm high with the white calcite stained by iron oxides, from the limestones of Ravenstonedale.

Abundant crinoid fossils that are each leg sections composed of a group of ossicles still attached to each other. These are in the limestone of Orton Scar.

sea swept in mainly from the lower north side, while the dark and muddy Kilnsey Limestone was deposited in wide bays on the south side. Only later (in Holkerian times) was the highest part of the block, around the sites of Ingleborough and Pen-y-ghent, finally submerged, under a long-lasting shelf sea. This was where the Cove and Gordale limestones, the two main elements of the Great Scar, were deposited across the entire Askrigg Block (though they are known by different names in some areas).

At this time (in the Asbian), the Askrigg Block was a classic carbonate platform. Corals, bivalves, bryozoans, crinoids and a host of smaller creatures thrived in the warm water no more than a few metres deep. Their calcite shells were broken down and reworked to create a sediment of carbonate sand, silt and mud that was then recrystallized into the strong limestone. Crowded groups of brachiopods left great banks of their thick clam-like shells, and various corals formed scattered patch reefs. Perhaps most importantly, lagoons were inhabited by great forests of crinoids; these were peculiar animals related to sea urchins but standing high on one leg like the stem of a plant, and washer-shaped segments (known as ossicles) of these cylindrical calcite legs are abundant in some beds. These all contributed to variation across the platform, but good fossils only survive at some levels.

The platform subsided by about 200m (650ft), to keep pace with the limestone accumulation, but the subsidence was slightly irregular. This created cycles within the limestone, spanning intervals with slightly deeper water being slowly lost to sediment accumulation, between each phase of subsidence. Varying sea levels complicated the picture, and also created short periods when the land emerged from the sea, with modest subaerial erosion and some accumulation of plant debris to form very thin coals. A scatter of shale layers (each under a metre thick) break

The Askrigg Block 340 million years ago – a shelf sea where limestone was being deposited in shallow lagoonal water, with patches of coral and a shell-sand beach. The drop-off into darker and deeper water matches the Craven Basin, but there were no hilly islands in the area now occupied by the Dales. This comparison view is actually of the shelf sea around Fiji.

the limestone succession, but are rarely seen except in cave walls. In part, they may represent clouds of land-derived mud, but their remarkable uniformity over great distances and their common occurrence on top of erosion surfaces suggest they may be largely ash from distant volcanoes (whose locations remain uncertain).

An edge to a carbonate platform, an environment where limestone forms, is buffeted by

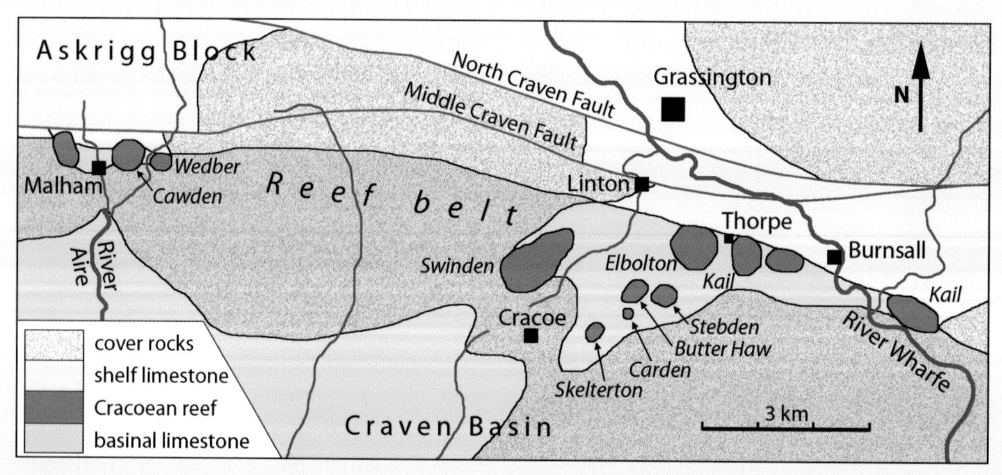

Distribution of the reef knolls that are features of the modern landscape around Cracoe and Malham. These formed in the deeper water of the Craven Basin, just south of the Craven Fault zone, which marked the edge of the shelf sea on the Askrigg Block. Reef limestone also extends between some of the knolls within the two groups. The cover rocks are Yoredale shales, Grassington Grit and Bowland Shales.

waves from the adjacent deeper basins, and the more oxygenated waters are ideal for growth of coral and algae, to form the basis of fringing reefs. The southern edge of the Askrigg Block, along the line of the Craven Faults, created this environment, now reflected in the reef knolls at Malham and in Lower Wharfedale. These knolls are made of a limestone that is notoriously structureless and appears to be largely a calcareous mud created by enhanced microbial activity and then quickly lithified. They are classic apron reefs, with lower ramparts of debris draped over the slopes down into deeper water. Corals, algae and bryozoans contributed to making a stable framework in parts of the reef, but the side slopes were ramps of fine-grained debris of broken shells and coral. They are sometimes known as Cracoean reefs, named after Cracoe, a village almost overlooked by the cluster of splendid reef knolls that form the isolated and rounded hills in Lower Wharfedale. This name distinguishes them from Waulsortian reefs, which are purely carbonate mud mounds formed as patches on the shallow marine platform; these form a variety of reef knoll not seen in the Dales.

The carbonate platform did continue both east and west of the Askrigg Block, but most of its limestones are not now seen. To the west, most have been lost to erosion beyond the Dent Fault, on the Lake District uplift, while to the east, away from the Pennine anticline, they lie buried far beneath the Vale of York. All this changed when (in early Brigantian times) increased inputs of clastic sediment and deeper cycles of subsidence created the Yoredale rocks of interbedded sandstones, limestones and shales – heralding the arrival of the great northern delta (*see* page 42).

A dusting of snow picks out the limestone reef knolls of Stebden Hill and Butter Haw Hill, two of the group that stand above Cracoe village; Grassington Grit edges Burnsall Moor behind the reefs.

A slab of the Girvanella Band, with the distinctive almond-sized algal nodules of the alga *Osagia* in a dark muddy limestone, from an exposure in Pen-y-ghent Gill.

Limestone is seen immediately east of the Dent Fault, where it forms the magnificent limestone pavements of the Clouds, high on the northern benches of Wild Boar Fell; it does also continue north onto the Alston Block, but only as a much thinner bed known as the Melmerby Scar Limestone.

The thick white limestone of the Yorkshire Dales owes its origins to deposition on the submarine platform of the Askrigg Block (*see* page 53). The southern margin of the Great Scar Limestone was therefore marked by the Craven Faults along the edge of the platform, where reefs of coral and algae grew overlooking the deeper water. These survive as the splendid reef knolls of Malham and Wharfedale; their steep grass-covered hills are dramatic features of the landscape, but their internal structure is a remarkably featureless limestone.

Basal conglomerate

The Great Scar Limestone was formed in a shallow sea where there had been only land for the previous fifty million years. Every piece of ground was therefore a beach environment for some short spell of time when the sea swept in over the subsiding landmass. The coarse-grained sediments of those beaches have been turned into rocks that still contain pebbles, cobbles and boulders. These are the basal conglomerate of the Carboniferous, sitting directly on the eroded stumps of older rocks, and overlain by the limestone formed in clear water after the beaches were further submerged.

Diagrammatic section through the main limestone units and related Carboniferous rocks across the Askrigg Block.

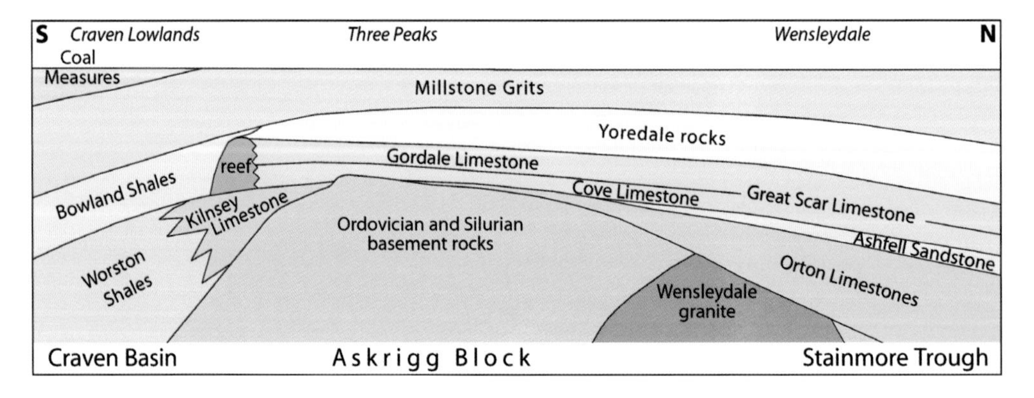

The pebble-bed conglomerate that lies at the base of the Great Scar Limestone, exposed below Norber Scar, just above Austwick. Scale is given by the ten pence coin.

Thornton Force, one of the Ingleton waterfalls, creates the classic exposure of the conglomerate forming an angular unconformity, with near horizontal Carboniferous rocks lying at odds with the older, more steeply inclined rocks below (*see* page 78). Rounded boulders, some nearly a metre across and made of the basement grit, form a conglomerate bed little more than a metre thick, but such a coarse boulder-run was probably a localized feature on the Carboniferous foreshore. At many exposures of the basal unconformity around Ingleborough, there is no conglomerate, because the sea swept in over a rocky foreshore that had previously been land. This surface was not flat, as the unconformity can be seen to vary in its stratigraphical level around Ingleborough. White Scar Cave is right on the unconformity surface at its exit in Chapel-le-Dale, and the stream passage descends through higher limestone beds, but, half a kilometre inside, it just cuts the top of a greywacke hill that was over 5m (16ft) high before it too was submerged.

The base of the limestone is well exposed at Nappa Scar, on Ingleborough. It lies just below the popular footpath that leads up to the Norber erratics, beneath a splendid basal conglomerate that was formed as a very well-sorted pebble beach. The footpath is along a ledge cut into a second, coarser, conglomerate (sometimes thought to be the basal bed where a very large block of slate at path level is easily mistaken for the basement). The angular slate debris and the poor sorting of this rock indicate that it is a debris flow deposit created by a Carboniferous landslide into the sea. It is not right at the base of the limestone, but it must have originated from a nearby hill even higher than the one in White Scar Cave.

Another coarse conglomerate, with well-rounded cobbles is seen in some small

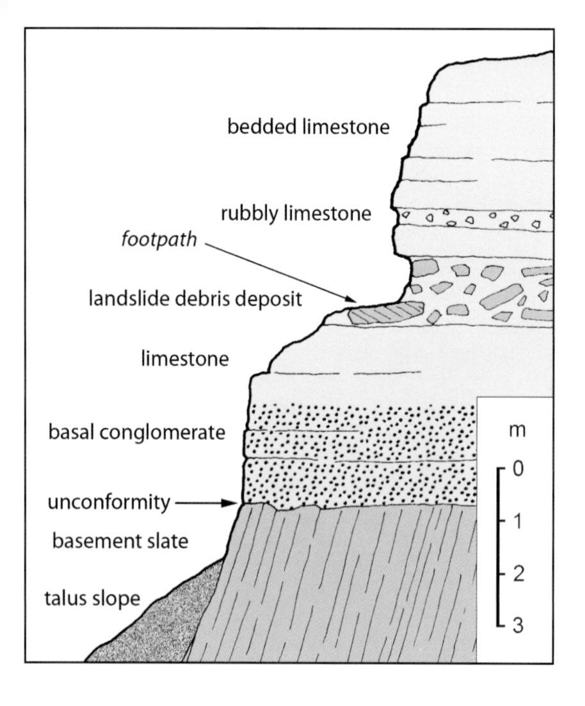

Section through the beds at the unconformable base of the Great Scar Limestone exposed beside the footpath along Nappa Scar to the Norber erratics on southeastern Ingleborough.

riverbed exposures immediately west of the Dent Fault. Sometimes known as the Sedbergh Conglomerate, its isolated position means that it cannot be placed in context with the other basal beds of the Carboniferous, and may even be of late Devonian age.

Limestones above the Great Scar

White limestone does create the character of the southern Dales, where the Great Scar is both at its thickest and is also most dissected – so that it creates the high karst plateaux around Ingleborough and eastwards to Wharfedale. But this is not the only Dales limestone. There are many more limestone beds within the succeeding Yoredale group of rocks (*see* page 34). They lack the big white cliffs, but are marked by some splendid karst features, contain some long caves, form a few significant cliff lines within the grit country of the northern Dales, and also contain the best of the limestone fossils. They should not be forgotten.

Limestones of the adjacent basins

South of the Craven Faults, there is no Great Scar Limestone, because there was no shelf sea in which it could form. Instead here was the deep Craven Basin, which slowly filled with sediments, dominated by the Worston Shales. These sediments do include significant beds of limestone, including the Chatburn Limestone, and these form Haw Bank and other hills near Skipton – except where they have been quarried away. They also include some thin beds

The coarse Sedbergh Conglomerate exposed beside the Rawthey River, below the road bridge just east of Sedbergh.

Evening light on a typical low scar formed by a strong bed in the upper part of the Great Scar Limestone, with the dark summit of Pen-y-ghent just visible. A drystone wall is made of the same limestone.

known as limestone conglomerate; these appear to be nodular beds, formed where calcite grew around shells when the bed was forming in the inter-tidal zone; these contrast with the debris origins of typical conglomerates. Overall, the basinal limestones are mostly very black, slightly impure and thinly bedded, so they are not strong enough en masse to form cliffs, scars or significant karst; they are very much overshadowed by the massive white limestones that lie further north in the heart of the Dales country.

West of the Dent Fault, and north of the Lake District block, the Ravenstonedale Gulf was, in early Carboniferous times, another deeper basin where mixed sequences of sediments accumulated. The Orton Group includes various shales and impure limestones, interbedded with the Ashfell Sandstone. Only later was the Gulf largely filled, and limestones comparable to the Great Scar were formed; these now form the low escarpments around Asby Scar that are noted for their extensive limestone pavements.

A steeply dipping bed of nodular limestone within the Craven Basin sequence, exposed in the wall of the old Hambleton Quarry, near Bolton Abbey.

CHAPTER 3

Banded Yoredales

Above the Great Scar Limestone, a mixed series of alternating beds of strong and weak rock is distinguished by the terraced slopes that are so common across the Yorkshire Dales. These rocks form the pyramids of Ingleborough and Pen-y-ghent (among others), each sitting atop the great platforms of Great Scar Limestone, and compose practically the entire slopes of Wensleydale and Swaledale. The stratigraphical age of these Yoredale rocks is largely Brigantian, but they also extend up into the Namurian, ending at the unconformity at the base of the Grassington Grit, which marks the beginning of the Millstone Grits (*see* page 40). The Brigantian Yoredales were once known as the Wensleydale Group; so they have borrowed both the old and the new names for Wensleydale (which is traversed by the River Ure, so is nearly 'Uredale').

Hillside benches above Cray, Wharfedale, are the hallmark of land on banded Yoredale rocks.

Yoredale sequences

The Yoredale rocks constitute one of Britain's finest examples of cyclic deposition; they extend through a succession about 150m (500ft) thick in the southern Dales to over

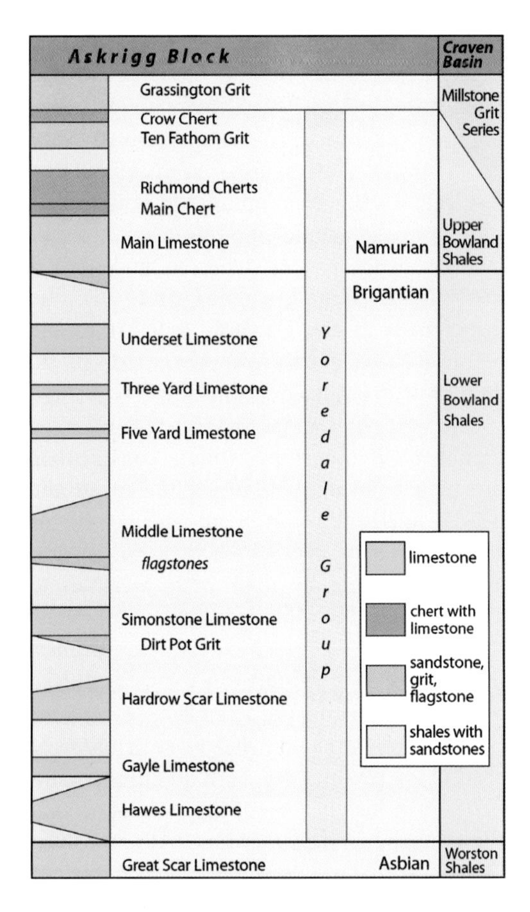

The main units of the Yoredale rocks in the Yorkshire Dales, and their equivalents on the Craven Basin. Individual bed thicknesses are very approximate because they are so laterally variable; the total thickness from the Hawes Limestone to the Crow Chert is around 300m (1000ft) in most areas.

400m (1300ft) thick around Stainmore. Beds of sandstone, limestone and shale are repeated in a total of eleven cycles called cyclothems, though all eleven cannot be seen in any one locality. It is correct to describe the first of these rocks as sandstone, because it was once mostly quartz sand that has since been lithified into strong rock, by the deposition of minerals in the pore spaces between the sand grains. But grit is the widely-used, colloquial term for a strong, generally coarse-grained sandstone that forms natural scars; grit is now an established part of Yorkshire language.

Within the Dales, Yoredale rocks form the higher parts of the Three Peaks – but not their summit caps. They also form the main slopes of the northern Dales, both Wensleydale and Swaledale, though the intervening moors are all founded on higher beds. The basic Yoredale cyclothem has a limestone, capped by a shale; this grades upwards with the grain size becoming steadily coarser, until it becomes solid sandstone. This, in turn, is generally capped by the next limestone. The whole sequence was formed in the great unstable delta environment of the time (*see* page 42). Limestones were formed in shallow water when the sea extended across the entire Askrigg Block, but these interludes lay only between phases of deltaic deposition of sand and mud – that eventually formed the sandstones and shales. A thin and discontinuous coal seam occurs within some of the sandstones near the top of the cycle, where a short-lived delta-top swamp developed, before it was buried by sediment as a consequence of renewed subsidence.

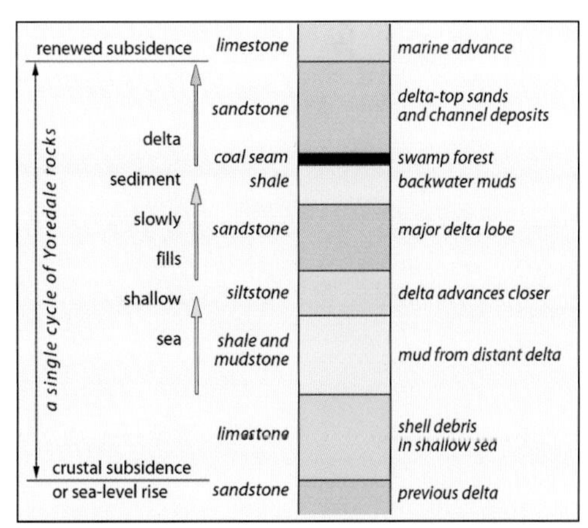

A typical Yoredale cyclothem, the sequence of rocks within a single cycle of deposition between two stages of crustal subsidence or sea-level rise. Not all cycles have the same sequence or thicknesses, and an individual cycle may vary between 10–60m (30–200ft) thick.

Hardraw Force

Alternating limestones and shales create some of the finest sites for waterfalls, where a strong limestone lip overhangs a plunge pool scoured out of soft shale. So the Yoredale beds are perfect, and Wensleydale is at the heart of waterfall country. Tallest of the Dales' cascades is Hardraw Force, just behind the Green Dragon Inn in Hardraw village.

A subdued terrace within the northern slope into Wensleydale hardly warrants description as a scar, but it is formed by the outcrop of a Yoredale limestone. This is known to geologists as the Hardrow Scar Limestone, because they stick to the older name of Hardrow – even though the village and waterfall have evolved into the Hardraw spelling. Hearne Beck and Fossdale Gill drain the slopes of Great Shunner Fell and join to form Hardraw Beck. On the way into Wensleydale, this is deflected east by a tail of drift deposited during the Ice Age (which breaks up eastwards into poorly shaped drumlins along the flank of the main dale). It then crosses a small fault and just catches the outcrop of the Simonstone Limestone, before flowing into a small ravine cut into the underlying shale and sandstone. It has entrenched only a metre or so into the upper beds of the Hardrow Scar Limestone before it feeds out to the waterfall, dropping 28m (90ft) from its limestone lip.

The waterfall cliff is capped by 7m (23ft) of dark bedded limestone, with a conspicuous overhang in an alcove to the right. Beneath the nearly horizontal limestone, 4m (13ft) of sandstones have a massive bed at the top, and are recognizable by their rusty weathering. These grade downwards in typical Yoredale style, become progressively finer sandstones, then siltstones then mudstones within a thickness of about 4m (13ft). Beneath these, 13m (43ft) of laminated shales are undercut around the plunge pool. Perhaps another 10m (33ft) of shales lie out of sight, and beneath these sits the Gayle Limestone, with the Green Dragon Inn standing on its outcrop further down the valley. Hardraw Force may be a splendid waterfall, but it also provides a fine exposure of most of one cyclothem of Yoredale rocks.

In front of Hardraw Force, a rocky gorge has been formed by classic waterfall retreat. Continuous scouring of the weak shales by the falling water has been, and still is, followed by intermittent collapse of the undermined caprock of limestone and sandstone. Hence the waterfall retreats, while retaining its vertical profile. Cliffs along the lower part of the gorge are weathered back so that its profile is more of a steep-sided valley, but the total length of gorge and valley is about 250m (800ft). At the downstream end the valley

Hardraw Force in dry weather, with the stream falling clear from the lip of limestone into the plunge pool scoured out of shale.

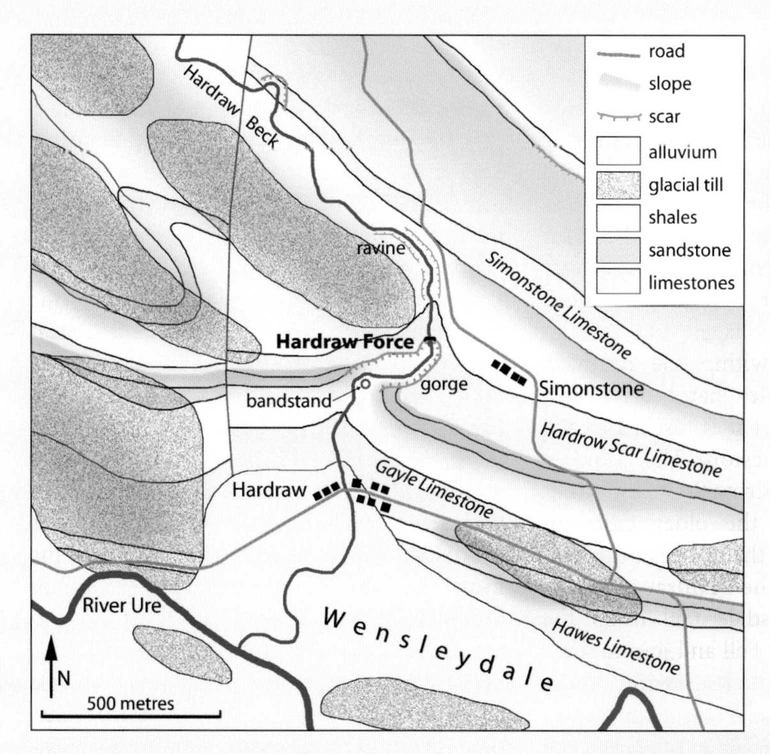

Outline map of the Hardraw area, with Hardraw Force at the head of its retreat gorge cut into the bench formed by the Hardrow Scar Limestone. Glacial till is only shown where it thickens into the drumlin hills and ridges.

breaks out through the limestone scar that lies along Wensleydale, and this was the site of the initial waterfall, established as this dale's glacier declined at the end of the Ice Age. So 250m (800ft) of retreat has taken about 13,000 years; Hardraw is one of those relatively rare sites where an actual rate can be ascribed to erosion processes, and it becomes a little more possible to understand the scale at which a landscape evolves.

Since 1884, the lower end of the narrow valley at Hardraw has been the venue for brass band competitions that are now an annual event. Steep hillsides create a natural arena, and the bandstand is round the corner from the waterfall, just out of earshot of its incessant rumbling. At the end of a day's competition within the valley, all the bands combine for a finale; Pachelbel's Canon played by two hundred musicians in the open air is unforgettable, and adds a special touch of magic to this corner of the Yorkshire Dales.

Brass band competition at Hardraw, with the waterfall just round the corner deeply shadowed among the trees.

Yoredale limestones

Within the Yoredales, about a quarter of the rocks are limestones. These were formed in short intervals that had clear Carboniferous seas, before they were swamped by deltaic sediment in each repetitive cycle. Between the Great Scar Limestone and the Grassington Grit, there are nine significant limestone beds, each about 5–25m (15–80ft) thick, that can be traced around the flanks of the northern Dales.

These limestones create their own distinctive landforms. Most of the well-defined rock scars along the flanks of the northern dales are formed by the limestones, which are generally the most erosion-resistant beds within the banded Yoredale sequences. For the same reason, they form the lips of many of the waterfalls. The Hawes and Gayle limestones create the Aysgarth Falls where the River Ure cascades over their outcrops, and the latter is named after the lovely stepped cascades on the beck down through Gayle village. Hardraw Scar is more of a grassy bench than a rocky scar in upper Wensleydale, but it gives the name to its limestone, which is best seen at Hardraw Force (*see* page 32). This splendid waterfall has a free drop because the stronger limestone (and the sandstone immediately beneath it) forms an overhang above the softer shales. Higher in the Yoredale sequence, the Middle Limestone becomes thicker and more important farther south, where it houses numerous stream sinks and potholes above Wharfedale, including the long cave passages behind Mossdale Scar, and also takes the River Nidd underground beneath the village of Lofthouse.

Lower Aysgarth Falls form a watery staircase over well-bedded limestones at the base of the Yoredale succession.

Mossdale Scar is made by the Yoredale Middle Limestone, where a stream on Grassington Moor sinks into a long cave.

At the base of the Namurian Yoredales, the Main Limestone is only about 10m (30ft) where it forms the distinctive lines of white scars one step down from the grit summit of Ingleborough and just below the more substantial dark grit cap on Pen-y-ghent. Farther north, it more than doubles in thickness, and contains the best of the karst features in the northern Dales. Oxnup Scar and Satron High Walls are formed by it on the south side of upper Swaledale, as are Whitfield and Ellerkin Scars on the north side of Wensleydale. The Buttertubs potholes, the short but dramatic gorges of Hell Gill and others like it, and also the God's Bridge caves in the floor of the Stainmore Pass are all in this one limestone (which is known as the Great Limestone on the Alston Block, north of Stainmore).

The thin Yoredale limestones are commonly much more fossiliferous that the rather barren Great Scar. Scattered patches within the limestones are packed with fossils – shells that were not broken into small pieces, as happened to the majority that just form the tiny fragments in the bulk of the rock. Some of the brachiopods survive well because of their thick shells, and whole banks of *Productus* and *Gigantoproductus* shells can be seen in some outcrops. Rugose corals, so-called because of their rough outer surfaces, survive locally as mini-reefs in position of growth, notably of *Siphonodendron junceum* (perhaps better known by its old name, *Lithostrotion*) along with rarer colonies of the beautifully ribbed *Actinocyathus floriformis* (again with an old well-known name, *Lonsdaleia*). The large, isolated 'horn' coral *Dibunophyllum* may also be found. A bed of dark limestone within the Great Limestone, just above the Yoredales, is rich in rolled individuals of this coral that were broken up by storms and buried in lime sediment, and forms the distinctive and beautiful Frosterley Marble further north in Weardale.

In the upper reaches of Swaledale, Wain Wath Force drops over the thin Underset Limestone and is overlooked by a white cliff of Main Limestone, both within the Yoredale succession.

Crinoids grew in such large numbers in the Carboniferous lagoons that the stems of *Actinocrinus*, and the calcite rings that they break into, contribute much to the bulk of some limestones. Crinoid stems are perhaps best seen in the Dent Marble; this is not a true marble, but is just a dense limestone that can take a polish (so that it is defined as a marble within the stone trade). The name was given to crinoid-rich beds in any of the Hardrow Scar, Simonstone or Underset Limestones found in the higher slopes of Dentdale.

A pair of the thick-shelled brachiopods, *Productus*, that cluster in banks in some limestones in the Yoredale Group. These are each about 50mm (2in) across.

Sandstones, shales, coals and cherts

Most of the Yoredale sandstone beds are only a few metres thick, and their topographic impact is minimized because most of them lie directly below stronger limestone caps. The best exposures are in some of the smaller waterfalls of the northern Dales, where the

Top view of a reef of the coral *Siphonodendron junceum* exposed in the bank of a small stream above Pen-y-ghent Gill; this piece is nearly 300mm (12in) across.

brown-stained blocky sandstone can be seen between the lips of limestone and the plunge pools in shale. Few of these Yoredale beds are really worthy of being called a grit, and some are so fine that the grains are barely visible to the naked eye and they can look very like the limestones. Others are well developed as flagstones. These are sandstones that naturally split, along bedding planes rich in flakes of mica, into beds about 50mm (2in) thick that can be used for roofing or paving. Most notable are those that have been mined on both sides of Wensleydale around Hawes (*see* page 189). Just to confuse the issue further, the thinnest flagstones are known locally as slates, merely because they can be used for roofing; they are certainly not slates in the geological sense of a metamorphic rock.

Shales form a significant part of the Yoredale succession. Together with some slightly coarser siltstones, these are the weak materials that have been eroded off the stronger limestones and sandstones to form the rock benches along the flanks of Wensleydale. Shale is rarely seen at the surface because weathering breaks it down to clay soils and

BELOW: The delicate calcite lattice that links the individual corals in a colony of *Actinocyathus floriformis*, from the Hardrow Scar Limestone on the flank of Fountains Fell. Each coral is about 15mm (0.5in) across.

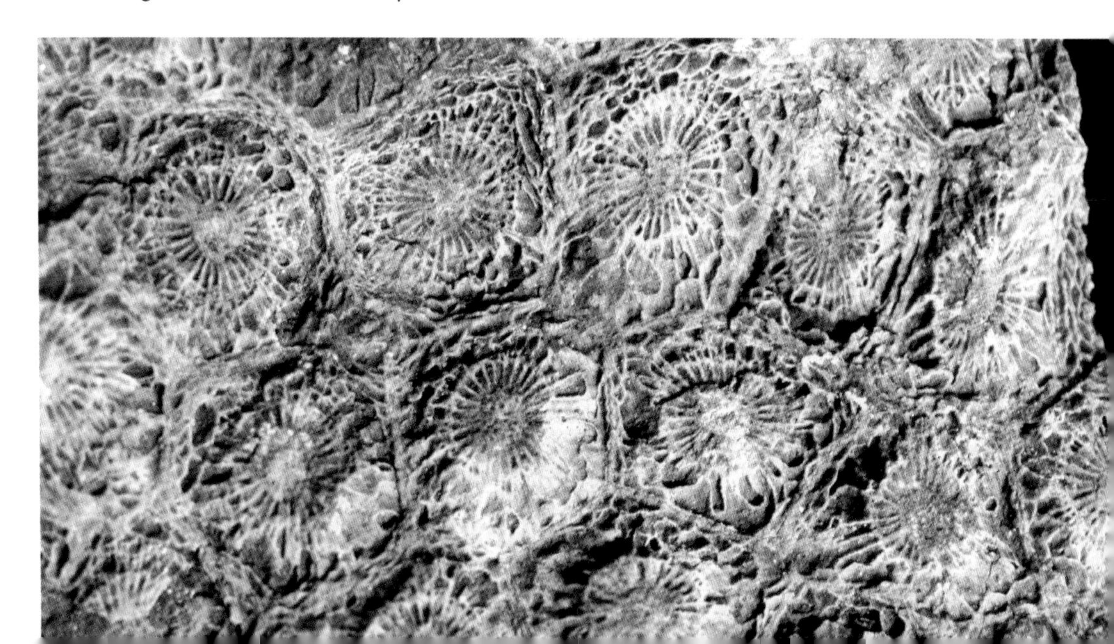

A Yoredale cycle exposed at Cotter Force in upper Wensleydale. The upper waterfall and the paler scars on the left are in the Hardrow Scar Limestone, while the lower cascade is over the underlying sandstone, into a plunge pool scoured out of weaker shales beneath.

mud, but there are outcrops in some of the stream beds, notably around the plunge pools of Hardraw Force and many other Wensleydale waterfalls (*see* page 32). Much of the shale is often described as mudstone. This is correct, because much of the rock is quite massive with poorly defined bedding. But it does weather by breaking down along closely-spaced and unseen bedding planes; so the rock exposed at outcrops is commonly well laminated, and has the appearance of a shale.

Deltaic conditions, which dominated through Yoredale times, allowed sporadic development of forests and swamps – some of which survived to create coal seams. There are two notable seams within the Yoredale sequence, both lying within sandstone beds that were formed on delta tops. One seam occurs just below the Middle Limestone along the western edge of the Dales, while a second lies just below the Main Limestone and was worked above Garsdale and elsewhere.

Another component of this group of rocks is chert. This is a really hard, very fine-grained rock, composed of pure silica, and very much like flint; much of it is pure black, but some beds are grey or cream-coloured on freshly broken surfaces. Even though it is so hard, most chert beds are well fractured, so they are less conspicuous in the topography than are the limestones. Between the Main Limestone and the Grassington Grit, a unit of chert up to 40m (130ft) thick is exposed in some of the higher slopes above Swaledale and eastern

Wensleydale – most notably capping the cliffs of Fremington Edge high above the lower end of Arkengarthdale. Its lower part is known as the Main Chert (and was an important host for the mineral veins), while its upper part is the Richmond Chert, the two being locally separated by mudstones and the metre-thick Little Limestone. Thick beds of pure chert are not common, as both parts consist of very mixed, banded sequences of chert and interbedded limestone with minor mudstone. The chert was formed when silica replaced the carbonate in entire beds of limestone, though some large brachiopod fossils still survive in the chert; the replacement probably occurred in the sediments of an almost enclosed lagoon, but there is some debate over the source of the silica, there being much more than might be expected from the sponges and planktonic radiolaria that had silica skeletons in the Carboniferous seas.

South of the Craven Faults, so within the Craven Basin, the lateral equivalents of the Yoredale rocks are found in the Lower Bowland Shales. These follow above the Worston Shales, but lack any interbedded limestone, having instead grit beds that increase in proportion upwards through the succession. As a group, these rocks are still weak and easily eroded, so that they now form the lower ground of the Craven Lowlands. It is only farther south that their stronger grits lie at outcrop to form Pendle Hill and other bits of high ground.

The mixed sequence of chert and limestone that forms the Richmond Cherts, exposed in the old Ever Bank Quarry near Middleham in lower Wensleydale. An erosion surface, clearly visible near the top of the face, marks where a lower bed was uplifted, tilted and partly eroded away in the shallow water environment before the next bed was deposited.

CHAPTER 4

Dark Grit

Strong, dark grit creates upland landscapes that are the perfect contrast to the white, scarred terrains of limestone. In the Pennines, the grit is never alone, but is mixed in with beds of weak shale, so that some of its landscape is broadly terraced; but its most characteristic landscapes are the huge expanses of moorland high up between the northern Dales.

The great succession of rocks that are dominated by grit and shale in the northern Yorkshire Dales are still of Carboniferous age. The thick beds of grit are known collectively as the Millstone Grits, and the oldest of them followed on from the Yoredale sequences about 325 million years ago, while the youngest that remains on the Askrigg Block formed about 10 million years later. Deposition was not consistent throughout that period, so the grits are interbedded with thick units of shale, and there are considerable lateral variations where the individual grits vary in thickness.

Millstone Grits

Key features of Pennine geology are the Millstone Grits. These lie midway through the

Strong, blocky grit overlying weak laminated shales are typical of the Millstone Grit succession; these are still exposed in the old quarry in Woodside Scar above the Scar House Dam in Nidderdale.

Millstone Grit is worked as building stone from the Hilltop Quarry at the head of Swaledale. This freshly broken block shows the *Liesegang* rings formed by bands of coloured iron oxides that are some of the cementing mineral between the sand grains within the rock.

Carboniferous sequence, above the limestone and below the Coal Measures. Though this group of rocks is dominated by shale, it is named as the Millstone Grit Series after its beds of strong grit, many of which were quarried to make millstones (though the industry was mainly based further south in the Derbyshire Pennines). These grits are coarse sand rocks composed of mainly quartz grains of varying size (known to geologists as 'poorly-sorted'), but which also include up to 20 per cent feldspar grains. As with the Yoredales, these rocks can equally well be described as sandstone or grit. The implication is that grit is a stronger variety of sandstone, but there are no clear definitions. Between them, the grits and the shales represent the climax of sediment deposition in the great Carboniferous delta (*see* page 42). Beds about 400m (1300ft) thick represent the Millstone Grit Series across their wide outcrops in the northern and eastern sectors of the Askrigg Block.

Grit moorlands are a major characteristic of the Pennine landscapes – from the Scottish border south to Derbyshire. Together with the White Peak of Derbyshire, the Yorkshire Dales limestone outcrops are really only minor interruptions in a mountain chain of grits and shales. Those great expanses of high, bleak moorland round the head of Swaledale, and the narrower strips between each of the northern Dales, are the result of strong grits resisting eons of erosion and protecting the shales that are interbedded with them. There are nearly a dozen grit beds in the sequence, but they are difficult to tell apart as the beds are of very uniform rock generally lacking in fossils. Only some thin, but very extensive, bands of marine shales contain fossils that allow reliable correlation of the various beds between distant outcrops.

One grit does stand out by reason of its geography and its geology. The Grassington Grit extends further to the south and forms that great, empty, block of moorland between Wharfedale and Nidderdale. This bed of grit, up to 50m (160ft) thick, originated as an active lobe of the Carboniferous delta just after the southeastern corner of the Askrigg Block had been uplifted. As a result of that uplift, much of the Yoredale sequence was eroded away, so that it is now missing down the east side of Wharfedale, and the Grassington Grit lies directly on the Yoredale Middle Limestone around Greenhow. Farther west, it is notable for forming the distinctive flat summit of Ingleborough. It also forms the upper ramparts of the other members of the Three Peaks, Pen-y-ghent and Whernside, along with Fountains Fell, Great Whernside and most of the higher hills that rise above the limestone

Once a Giant Delta

Most of the rocks in the Yorkshire Dales are of Carboniferous age. Carboniferous means coal-bearing; coals form in deltas, where wetland is created on top of huge piles of sediment dumped at the mouth of a river that enters the sea. One huge delta was the big event in Carboniferous times.

In the Dales region, the early Carboniferous saw a shallow sea of clear water where the Great Scar Limestone was deposited. At the same time, there was an active delta farther north, growing outwards from a landmass to the north and west, across the Edinburgh region. Some land-derived sediment was swept into the Craven Basin and Stainmore Trough, but the Askrigg Block remained clear. But all deltas grow, and that delta that had a small start in Scotland grew and extended southwards, filling in the entire limestone sea, until it reached another low landmass south of a point where Birmingham now stands.

Within the Dales, atop the Askrigg Block, the first signs of that giant delta were the sediments of the Yoredale beds, best seen, of course, along the slopes of Wensleydale from whence came the geological name – as Yoredale is an old name of the valley carrying the River Ure. These Yoredale rocks are distinguished by their repetitive beds of sandstone, limestone and shale, which form the rock sequence between the Great Scar Limestone and the Millstone Grit within the Carboniferous succession.

Within each Yoredale cyclothem, the shales represent the mud carried farthest out from the active delta. As the mud slowly filled the shallow sea, the delta grew across it by the banks of river-borne sand continually building outwards, with new sand deposited on their slip-off slopes into deeper water. The area was then a delta flat, with sand being deposited in the channels, mud being dumped in the inter-channel backwaters and locally swamps establishing to create discontinuous coal seams. Channels changed in major flood events, and new delta lobes grew between the older ones, where the rivers found easier sites to dump their sediments. The complex geography of a delta is reflected in the way that the Yoredale rocks are seen to be slightly different at outcrops on opposite sides of a hill or dale.

Then a little extra subsidence allowed the sea to sweep back in, so that limestone was deposited for a short while until the delta grew back with another cycle of sediments. Without the subsidence, deposition would have been transferred to other lower areas. So a thick rock sequence is a product of subsidence, and the weight of the new sediments adds to the weight on the local crust, producing sag, which causes the surface area to subside. While each return of the sea is most likely due to a burst of accelerated subsidence, it could be due to a rise in sea level. Such repetitive rises could only be due to climatic oscillations, when water is trapped and released from ice

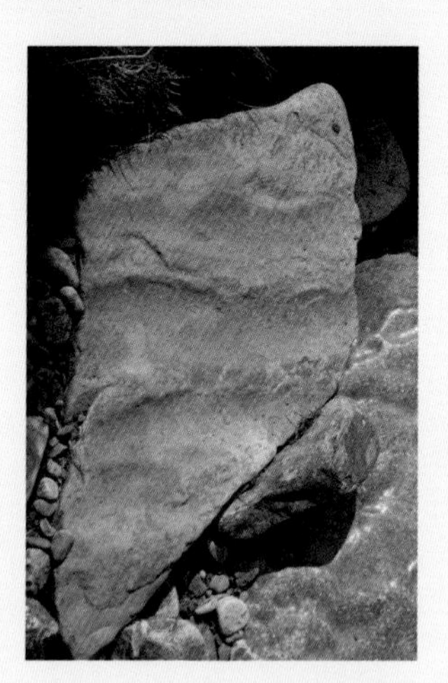

A loose block of Yoredale sandstone, found in Hell Gill, Mallerstang, with ripples formed when the sand was deposited in shallow water in the great Carboniferous delta.

One of many examples of cross-bedded Millstone Grit exposed on the tors of Brimham Rocks. These are the clearest signs of deltaic deposition, showing how the sand accumulated in sloping layers down the front of growing delta fans.

A delta now on top of a mountain – the deltaic Grassington Grit which forms the upper ramparts of Pen-y-ghent. The white cliffs below the Grit are formed of the Yoredale Main Limestone.

caps on distant continents. There certainly were Carboniferous glaciations, but interpretations of the Yoredale rocks currently favour just deltaic subsidence as the cause of their cyclic changes.

Eventually the giant delta grew down from the north until the Yorkshire Dales area was at the heart of the main deposition (*see* map on page 59). This was when the massive units of sand were deposited, to form the thick and extensive grit beds within the Millstone Grit sequence. The Grassington Grit was one such unit – a massive lobe of sand that grew out in a huge fan from the north and across the whole area now carved into the eastern Dales. And above and below it, thick shales and patchy coal seams represent the other phases of delta sedimentation.

Over time, this active delta spread south across the whole of what is now northern England. And once the sea was filled in, delta-top swamps could establish across huge areas. Britain then looked like the Mekong Delta does today. In the dense equatorial forests, dying trees fell into the semi-stagnant water where they could not oxidize away, and were then buried to form the coal seams. With intermittent subsidence and renewed delta growth between phases of wetland forest expansion, the Coal Measures were formed. They now underlie the industrial lands of Lancashire and Yorkshire, but were eroded away over the Pennine anticline, so that none survives within the Dales, save for that little patch on the fringes at Ingleton (*see* page 48).

This delta came from the north, and it is of course inconceivable that one this size could originate from Scotland. But back in Carboniferous times, there was no Atlantic Ocean. The river that created the delta drained off a huge chunk of what is now northern Canada, together with what there then was of Greenland and also Scandinavia. This was a river of truly continental scale. Its size compares very well with the modern Mississippi, which drains much of the same continent in the opposite direction. The geographies of today's Mississippi delta and the bygone Pennine delta are a remarkably good match in terms of size, sedimentation and environment.

The Pennine delta was the Millstone Grit, and it then led on to the Coal Measures. So perhaps we should thank the Canadians for their supply of sand, which built the delta, which made the coal swamps possible, which gave us the minerals of our industrial heritage, which supplied the growth of Victorian Britain, which gave us the foundations of our nation's economy and character. Without that Carboniferous delta, Britain would have been a much poorer country – and it could never have contained the landscapes of today's Yorkshire Dales.

The Askrigg Block 310 million years ago – a wetland of swamps and half drowned forests where plant debris was accumulating in the stagnant water to create the coal seams of today. This matches the area now occupied by the Dales, except that there is no large river to bury and preserve the plant debris under a sand delta. This comparison view is actually of the Okefenokee Swamp in Georgia, USA.

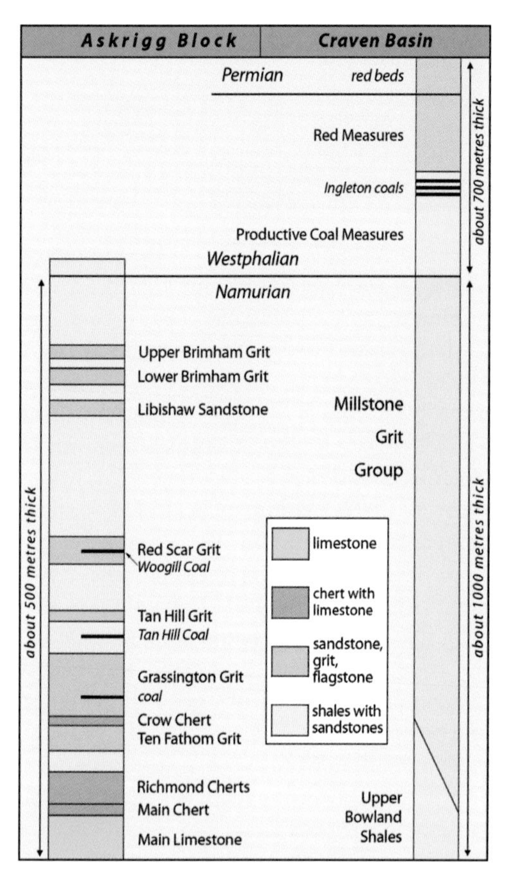

Notable beds within the Millstone Grit Group in the Yorkshire Dales, and the Coal Measures of the Ingleton Coalfield. Individual bed thicknesses are very approximate because they are so laterally variable; note the rough overall thicknesses shown in the margins.

benches of the southern Dales. In these western areas, the bed is sometimes known as the Lower Howgate Edge Grit, the bed of the same age that crops out round the head of Swaledale. On the Askrigg Block, the base of the Grassington Grit is something of a terrain marker, as it broadly separates the terraced slopes on the mixed Yoredale sequences from the bleaker moors on both the Grit and all higher beds.

In the southeast corner of the Dales, the best-known Millstone Grit outcrop forms Brimham Rocks (*see* page 46). Northwards from that wilderness of rock tors on the moor above Nidderdale, the Lower Brimham Grit

The open expanse of Grassington Moor typifies the rather bleak uplands that have developed on the wider and higher outcrops of the Millstone Grit.

An outcrop of thick shales between grit beds in the Millstone Grit succession, exposed by Mould Gill, close to Tan Hill on the moors north of Swaledale.

forms isolated crags as far north as the Masham area, but none rises to match its namesake.

Just as among the Yoredale outcrops, the thick beds of shale that lie between the various units of Millstone Grit are weathered to clays and soils. They are rarely seen on the high fells (as their outcrops are overshadowed by the grit crags), but their presence is often indicated by seriously wet ground on their leak-proof clay soils. There are also some thin coal beds, notably that which extends under the Tan Hill area north of Arkengarthdale. Nearly all the coal outcrops are marked by traces of old small-scale mining (*see* page 48). The Main Limestone and the Richmond Cherts are of Namurian age, but, along with a few shales beds are a continuation of the Yoredale rock sequences up as far as the base of the Grassington Grit.

Above the Yoredale limestones, the Dales grit country is hardly notable for its fossils. Most of the sandstones are fairly barren; few animals lived in those environments of active deposition and shells did not survive well in the coarse sand. Fossils are better preserved in the shales, but they are still not common. Most are in the thin marine bands, which contain goniatites, the contemporary style of spiral ammonoids, together with *Lingula* and a few other flimsy, flattened brachiopods, but none survives very well in weathered material.

Brimham Rocks

On a day of swirling low cloud, fantasies can be woven around half-seen glimpses of these looming rocks that have been fretted into ghostly profiles. Brimham Rocks form a group of tors – crags of bedrock that remain where weakened rock in between has been eroded away, mainly along large joints. They stand on the most easterly of the grit moors that overlook Nidderdale, just east of Pateley Bridge.

The Rocks are dramatic exposures of the Lower Brimham Grit, one of the thicker and stronger units of Millstone Grit. Originally a single unbroken slab of grit, its erosion into the numerous tors has created splendid sections exposing the internal rock structure. The Brimham Grit was formed as a complex series of sand banks, growing miniature delta fans and sand bars in river channels, which were all just parts of the giant Carboniferous delta. Details of the original beds of sand and pebbles can now be clearly seen on many of the tors. Most conspicuous is the current bedding that developed where sediment was deposited on underwater slip-off slopes, either on the fronts of advancing mini-deltas or on the outside edges of shifting sand bars. The individual beds sloped down at about 25 degrees into deeper water, but directions were inconsistent because of the complexity of the meandering channels within the grand delta area.

But it is the subsequent erosion of the grit that is so remarkable at Brimham. The Rocks are classic tors, but that just means that they are residual blocks of *in situ* rock ringed by steep faces. There is considerable debate as to why the Brimham tors are so much more spectacular than nearby grit outcrops. Tor formation requires two factors – good joints, and then good weathering of those joints.

It is not clear exactly why the Brimham Rocks are laced by these widely spaced fractures. The site is close to the eastern end of the Craven Fault zone, where major fractures could be expected, either as small faults or just joints (with no displacement on them). Then the joints may have been opened up a little as the blocks of grit moved slightly apart. This does happen along and close to escarpment edges, where the grit blocks gently slide out towards the unsupported valley side, or even sag into the underlying shale when it is softened by weathering and squeezed out into the same valley slope. Brimham Rocks are a fine example of scarp-edge tors, with their best examples almost overlooking the valley to the west.

Such blocks of grit on a scarp crest can also be pulled apart by glacial drag. This occurs where a creeping ice sheet or glacier is frozen onto the grit outcrops and drags them along a little by sliding them over a soft shale, which remains slippery below the zone of ground freezing. Pleistocene ice moving from the north could have contributed to heaving the grit blocks apart on Brimham Moor, though there is no solid evidence of its involvement.

The second phase of tor formation is the weathering that weakens the joint zones, so that the adjacent rock can be eroded out.

Just some of the splendid grit tors that make Brimham Rocks into a family playground for summer weekends.

Tors along the western edge of Brimham Rocks, where joints have opened up by a little sliding into the wide, glaciated valley of Nidderdale.

Again there is debate as to when this happened. It may be traced back to deep sub-surface chemical weathering under sub-tropical conditions during the Tertiary (perhaps ten million years ago), when water percolating down the open joints could produce bands of rotten rock – partly because the significant proportion of feldspar grains reacted with water to produce clays. Alternatively, intense frost-shattering during Pleistocene phases of periglacial activity could see the zones of jointed rock reduced to loose sand grains. Probably both processes were involved. Whether rotten or just loose, the zones of rock were easily eroded out by rainwash, wind or slumping, to leave between them the residual blocks that are the tors of today.

The added beauty of Brimham Rocks comes from the varied profiles that have been etched into the tors. This is due to selective weathering that has picked out the internal features of the rock. Weaker beds, probably those with more clay in them, and also individual bedding planes (many of which are themselves just very thin layers of clay), have been eaten away, commonly undercutting stronger beds. Such selective etching may be due to just weathering, either by reaction with water or by intense freeze-thaw activity during the Pleistocene. But it could also have been aided by wind erosion – by strong winds of cold air that rolled down off the Pennine ice cap. Known as katabatic winds, these are well known in glacial regions today. They would have blasted the rock tors with loose sand grains that were abundant in the cold, bare terrains of the Pleistocene – a process that is mainly effective within a metre of the general ground level, because wind rarely carries sand high into the air. Some of the tors do look like desert mushroom rocks, though Brimham's history relates more to Greenland than it does to the Sahara. And the various tor profiles of anvils and wedges are each shaped from the angles of the current bedding in the rock – yet more details of the landscape that originate in the underlying geology.

Idol Rock, a magnificent 'mushroom rock' at Brimham Rocks. The block of grit stands 6m (20ft) high and weighs around 200 tonnes, standing on a narrow base that has been undercut along a weaker grit bed, perhaps by wind erosion during the Ice Ages.

Search for Coal

Coal has long been a valuable commodity, whether it is in a thick seam under a major industrial city or in a thin seam high on the Pennine fells. The great sequences of Carboniferous shales and sandstones that overlie the Great Scar Limestone in the Yorkshire Dales are ideal sites to find coal; they fully live up to their stratigraphical name – they contain carbon – and thin seams of coal are scattered across almost all the high fells. None is workable in today's economy, but they have proved very welcome to local miners of the past.

Largest of the past mining concerns was the Ingleton coalfield, just south of the village. After coals had been found exposed along the banks of the River Greta, there were by 1650 a line of small bell pits between Ingleton and Burton, notably near Raygill. Each bell pit was a shaft down to the coal, which was then worked outwards until the roof became unsafe, when it was abandoned and another was sunk nearby. Later mines were mainly 30–60m (100–200ft) deep, and worked the Six Feet and Four Feet coals by traditional pillar-and-stall methods, leaving pillars of coal to support the roof. By 1835, annual production reached 16,000 tonnes, and forty years later there were seven working mines. In 1914 the New Ingleton Colliery, close to the village, was opened to work the coal down to depths of 250m (800ft). By good fortune, the shaft sinking revealed two more seams, each about 3m (10ft) thick, but these lay in only a small area and were worked out by 1918. The extent of all the coal seams was limited – by outcrop to the south, by the Craven Faults on the northwest, and by erosion under the Red Beds unconformity to the northeast. Mining ended in 1940, leaving only the deepest part of the Six Feet seam unworked, perhaps uneconomic under competition from the larger coalfields further south. Little remains of the old Ingleton mines beyond a few grassed-over tip-heaps, though a shaft cap at Wilson Wood collapsed in 2003 to create a large crater (which was soon sealed and filled).

A fossil tree trunk standing in position of growth in the Ingleton coalfield. Its soft core rotted away when it was buried in mud, and was then filled with sand to form this mould, before the stronger bark also rotted away. So now there is no coal left, and it stands in mudstones exposed along the bank of the River Greta.

Another little slice of Coal Measure rocks form what has generously been called the Stainmore coalfield. There are nineteen coal seams within a 250m (800ft) thick sequence of Coal Measures. But most seams are thin, and they dip at about 65 degrees, extending only short distances before being cut off by major faults, so their mining has been minimal; only a few collapsed adits remain. These coals were formed in the Stainmore Trough, but lie just west of the Dent Faults where they have been saved from erosion in the lee of the Askrigg Block.

Second only to Ingleton in terms of importance, a group of mines lie in Tan Hill, high on the bleak moor north of Swaledale and marked by the well known inn at the road junction 528m (1730ft) above sea level. Small drift mines and bell pits may have been started in the twelfth century, but the mines grew when they were supplying coal to the lead smelters not far to the south. The Tan Hill coal is a seam a metre thick midway through the Millstone Grit. Almost horizontal, it underlies large areas of the moor at accessible depths; Tan Hill Colliery, King's Pit and Mould Gill level were among the larger workings, and King's was the last to close, in 1934. The same coal was also mined until 1927 from collieries in Great Punchard Gill to supply the local

An old adit entrance to the coal mines of Tan Hill lies half buried behind collapsed shale in the steep bank of a gulley into Mould Gill.

smelt mills in upper Arkengarthdale. Smaller workings lie as far south as the slopes of Great Shunner Fell, but non-deposition in the laterally variable Carboniferous delta means that it is absent from the rock sequence in many areas.

Along the eastern side of the Dales, a number of coal seams occur in the Grassington Grit, where it splits into multiple beds of coarse sandstone with shales and coals in between. The Grit was formed as a complex deltaic unit, so the coals are discontinuous, each formed wherever a delta-top swamp was briefly established and then buried. The largest mines were under Threshfield Moor, west of Grassington, where one seam can be traced along about 3km (2 miles) of outcrop and reaches up to 0.9m (3ft) thick. Between 1607 and 1905, this was worked intermittently from at least three shafts, each up to 20m (65ft) deep, with a drainage level out to Grysdale Beck. Thin coal seams within the Grit were also worked on the summit of Fountains Fell, in the Aket Coalpits now just above the south shore of Grimwith Reservoir, at various sites in Nidderdale, above Kettlewell by Top Mere on the saddle over to Coverdale, and also high on Grinton Moor where a colliery worked from 1768 to 1895. Most of these supplied nearby smelters extracting metals from the vein ore (*see* page 169). Only grassed-over tip-heaps, either large fans below old adits or smaller rings around old bell pits, remain to create features that appear slightly unnatural in the modern landscape.

Stratigraphically higher than the Tan Hill Coal, and restricted to the eastern edge of the Dales, the Woogill Coal lies within the Red Scar Grit and was worked mainly in the slopes of Colsterdale. On the western side of the Dales, a discontinuous coal seam lies just below the Middle Limestone and has been worked in Coal Gill high on the slopes of Whernside and also in both Sleddale and Cotterdale, where it reaches nearly half a metre in thickness. The Garsdale coal pits were a productive cluster beside the fell road over into Dentdale; they worked yet another seam that also reaches 0.5m (18in) thick, commonly by shafts sunk through the Main Limestone which lies just above it; the same coal was also worked in the Cross Pits, high in Arten Gill. Coals any thinner than that could only be worked in the smallest of open diggings along hillsides, but many were in the days before better transport made coal more widely available.

An alternative fuel was provided by peat. Little of the upland peat can compare with the thick and rich peats of lowland bogs, but there were many areas where fuel could be won for the effort of a little digging. The eighteenth century saw extensive peat extraction on the high moors above Swaledale, when it fuelled the local lead smelters; the ruins of large peat stores remain above the Old Gang and Grinton smelters. Peat is often known as turf – perhaps a better description of a partially rotted organic soil that was also a cheap home fuel used by everyone in the past. Large numbers of householders held turbary rights – the right to cut peat from a given stretch of high fell purely for their own use. Turbary Road above Kingsdale, Ten End Peat Ground on the northern shoulder of Dodd Fell, Peat Lane up from Pateley Bridge and Peat Gate that climbs from Swaledale to the fell all tell of bygone workings of peat. But it is no longer a popular fuel, and regrowth has largely obscured the shallow fell-top diggings of yesteryear.

Beside the lonely road across the moors above Gardsale Head, a timber-capped shaft amid a patch of white limestone is all that remains of a mine that dug coal from just below the Yoredale Main Limestone.

Reticuloceras is one of the more common goniatite fossils that occur in the Millstone Grit succession; this specimen in black shale is about 25mm (1in) across.

Grits, shales and coal in the Craven Basin

South of the Craven Faults, and off the Askrigg Block, the rock sequence is very different in the Craven Basin, where there was continuing subsidence and thick sediment accumulation in the early Carboniferous. The Upper Bowland Shales are the Basin's lateral equivalent of the lowest Millstone Grits on the Block. They extend upwards into a sequence of grits, sandstones and shales that are broadly comparable to the Millstone Grits of the northern Dales, but are very different in their detailed stratigraphy; these stronger rocks create the higher ground of the Forest of Bowland. The softer terrain directly south of Giggleswick Scar is on a sequence more dominated by shales, so that the rolling farmland on good soils is so different from the limestone country – and is therefore not part of the National Park.

The Grassington Grit also reaches across the faults, and is known as the Brennand Grit farther south in the Craven Basin. It forms the rim of crags and much of the cap on Embsay Moor. On the rolling shoulder of Threshfield Moor, just south of the limestone Malham Moor, another outlier is distinguished by containing a thin coal seam that was mined in the past (*see* page 49).

Though just into the Craven Lowlands, the Ingleton coalfield very neatly rounds off the story of the Dales geology. This little patch of Coal Measure rocks is hidden away beneath the farmland of the Craven Lowlands immediately southwest of the South Craven Fault, where it passes through Ingleton village. About 400m (1300ft) of shales and mudstones contain five coal seams, which were formed just a few million years after the last of the Millstone Grits were deposited. All the seams have been mined, two of them quite extensively (*see* page 48), along with some clays suitable for making pottery; but the sparse bands of ironstone nodules were never economic. Coal originated as plant material, but was formed largely as a sludge of disintegrated debris, so plant fossils are better preserved as individuals within the adjacent shales. Fern fronds of *Neuropteris* are perhaps the most common, and sandstone casts of upright trees are periodically exposed within the eroding mudstone river banks upstream of Burton. Some bands of the shales have the non-marine bivalves, notably *Carbonicola*.

The productive Coal Measures are covered by another 400m (1300ft) or so of red mudstones and sandstones marking the dying stage of the Carboniferous, which are transitional to

Fronds, each 100mm (4in) long, of the fossil fern *Neuropteris* occur on just some horizons in the sandstones of the Coal Measures.

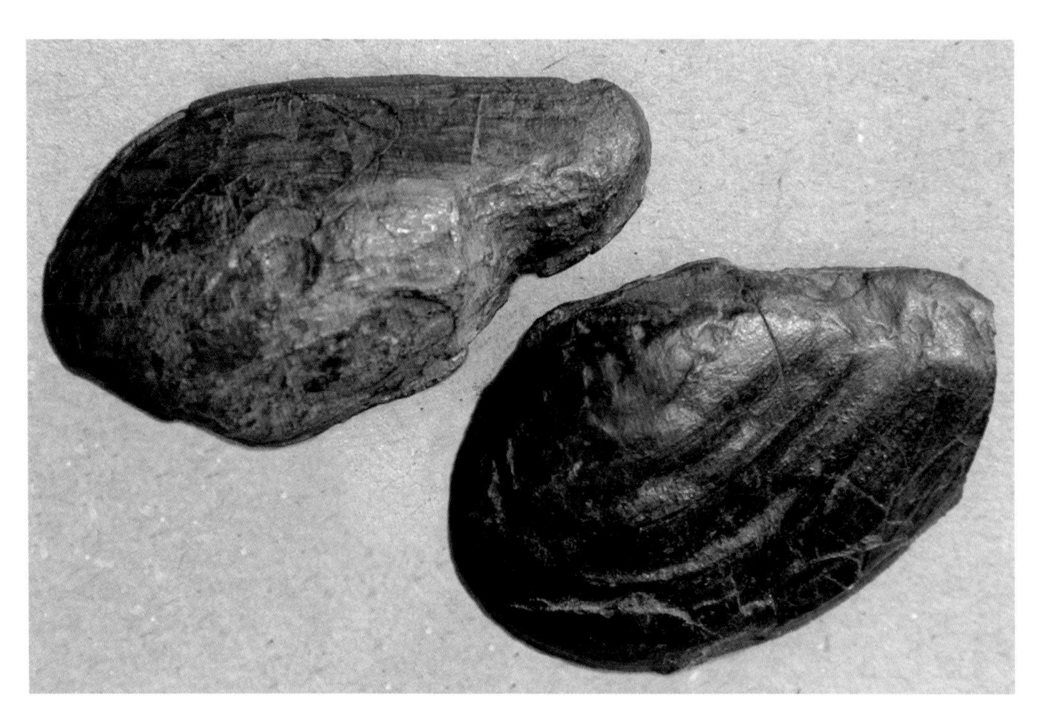

A pair of the non-marine bivalve *Carbonicola* that can be found in some of the shales in the Coal Measures of the Ingleton coalfield. Each is about 25mm (1in) long.

the desert red beds of the Permian and Triassic. A rather fine roadside outcrop is overlooked by many, as they trundle along the A65 just east of Cowan Bridge. This red-stained rock with rough limestone cobbles, known as a breccia, lies at the top of the succession; as it is devoid of fossils, it remains undated, but is probably from the earliest years of the Permian.

All traces of both the Coal Measures and the red beds have been removed by erosion from the uplands of the Askrigg Block. And no younger rocks survive within the Yorkshire Dales. The next materials to be seen in the Dales are the river sediments and glacial debris of the Quaternary – so young that they have not been turned into rock. But theirs is the story of landscape evolution (*see* page 66).

One of the few exposures of the red beds that overlie the Coal Measures in the Ingleton coalfield. These are in a cutting along the A65 road, and the red colour was only conspicuous in the rock freshly exposed by road widening some years ago.

CHAPTER 5

Rock Jigsaw

While details of a landscape are commonly determined by the individual rock types, the broad patterns and regional contrasts owe much to the way the different rocks fit together in the three-dimensional jigsaw that is the ground beneath our feet. Today's geological structure is the product of millions of years of evolution, where new sediments have accumulated and then been turned into rock, and where parts of these have been removed by erosion before the next rocks were formed.

Perhaps most importantly, what were once horizontal layers of undisturbed sedimentary rock have been contorted, crumpled and broken when caught between colliding continents in the processes that are now well known as plate tectonics.

One way to really understand the rock jigsaw of the Yorkshire Dales is to follow its evolution through geological time – and this does make a great story. A rather grand concept, or at least an acceptance, of scale is required. Not only have the rock units within the Yorkshire Dales moved by kilometres relative to each other, but the whole lot has travelled thousands of kilometres on the restless conveyor belt of the shifting plates and drifting continents.

Early events in the basement

The story of the Dales starts down near the South Pole, when sand and mud were deposited in the Iapetus Ocean, adjacent to Avalonia, a small slice of continental crust that had broken away from the Gondwana supercontinent. These were eventually to form the

The classic unconformity at the base of the Carboniferous, so well exposed in the old Combs Quarry at Foredale, in Ribblesdale. Great Scar Limestone is almost level where it lies across the eroded ends of steeply dipping beds of Horton Flags. That simple break between the two represents about 80 million years of geological time, when the flags were deeply buried, folded, uplifted, eroded and then submerged beneath the Carboniferous sea.

rocks of the Dales basement now exposed around Ingleton (*see* page 14).

Across the Iapetus Ocean lay Laurentia – the huge continental slab that is still exposed across much of northern Canada and underlies Greenland's icecap. As Avalonia and Laurentia closed in on each other the sediments and rocks of the intervening ocean floor were crumpled up; sandstones and shales were turned into greywackes sequences in yet another phase of plate convergence.

Avalonia and Laurentia finally collided and merged into one about 400 million years ago; the Iapetus Ocean was no more. Deep down, in the core of the collision zone, some rocks melted, and the Wensleydale granite was formed. At shallower depths, the basement rocks were crumpled into tight folds, creating the vertical beds that can be seen today in the Dales inliers around Ingleton and Horton. The ocean was gone, and the new Caledonian fold mountains rose far above sea level. Erosion got going, and within less than 50 million years the new landmass was worn down to nearly flat lowland. The almost level pre-Carboniferous unconformity, seen in outcrops and quarries around Ingleborough, is that ancient surface, formed across the eroded ends of even older beds of rock.

Another lasting consequence of the Caledonian collision of the continents was the establishment of the deep basement structures. Like two ice floes grinding into each other, long fractures reached far back into the continental slabs when they had to adjust their shapes to fit against each other. And those fractures became inherited weaknesses that defined the shapes and positions of many later geological structures.

Key events in the Carboniferous

Following the collisions and compression of the earlier plate convergence, the section of the Earth's crust that is now northern England was subjected to significant tension. Extension of the crust produces two notable effects. Firstly, the basement breaks along major faults, many of which are inclined at perhaps 30–45 degrees from the vertical. So adjacent blocks can pull apart by one sliding down the edge of the other. Blocks therefore end up at different

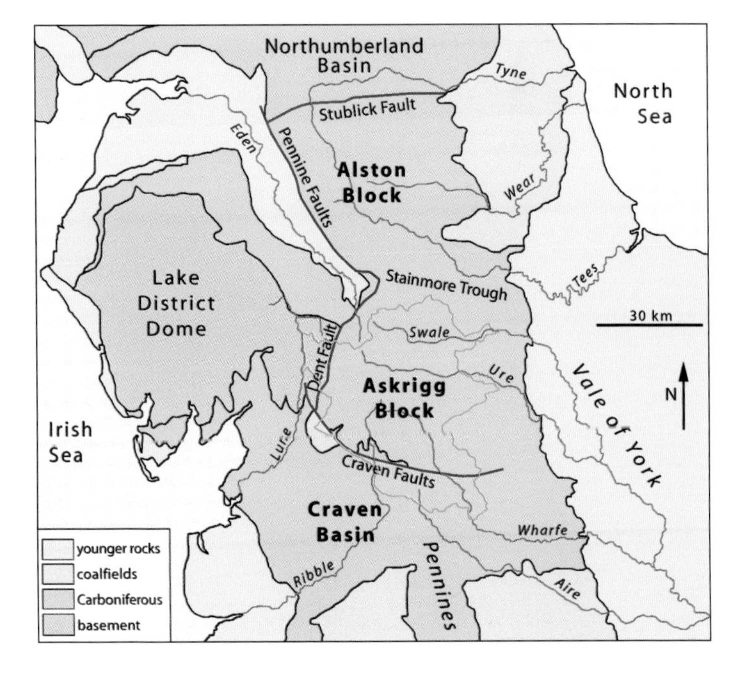

The position of the Askrigg Block within the geological structure of the Pennines and their surrounds; the boundary of the Yorkshire Dales National Park is picked out in dark green.

Craven Faults

The great valleys that are the Yorkshire Dales are all cut into the Craven Uplands, while to the south lie the contrasting Craven Lowlands. The common boundary is the line of the Craven Faults, along the southern margin of the Askrigg Block, on which stands the high ground of the Uplands.

These faults constitute a geological classic, with three major faults and a host of minor fractures all within a very visible fault zone. The North Craven Fault has the smallest displacement of the three but is the longest, as it can be traced from west of Ingleton to beyond Pateley Bridge. The South Craven Fault is only a few hundred metres from the North Craven near Ingleton, but eastwards it swings away to the south, and then breaks up into a splay of faults beneath the Craven Lowlands; its southern element is sometimes known as the Gargrave Fault. The Middle Craven Fault branches from the South Craven at Settle, and extends east as far as Wharfedale where it probably merges with branches of the North Craven.

Though the whole fault zone creates a major break in the Pennine landscape, the individual faults are rather lost in the details of the landscape along most of their traces. Giggleswick Scar is the most splendid fault feature, with its limestone crags rising west of Settle. But this is not a fault scarp – which is defined as a cliff formed by the fault displacement; instead, it is a fault-line scarp fault, created by differential erosion where strong limestone has been left higher than the weaker rocks south of the fault. Malham Cove could be described as a feature of the fault scarp, but the cliff that stands today has retreated about 600m (2000ft) from its origins on the fault (*see* page 106).

A rainbow stands on the edge of the Craven Uplands where the Great Scar Limestone forms Robin Proctor Scar, above Austwick. The North Craven Fault lies along the foot of the scar, separating the high white limestone from the green fields of the Craven Lowlands.

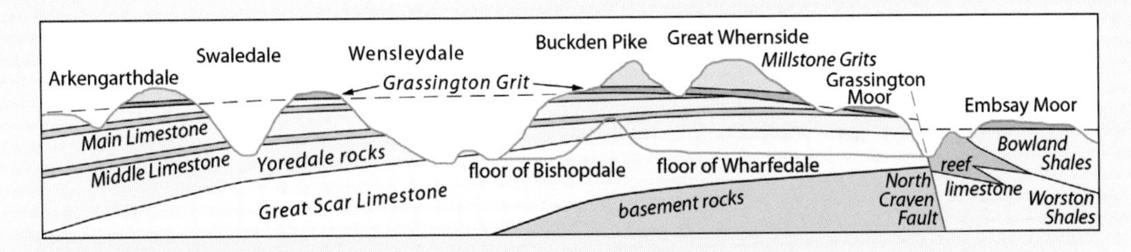

Long profile Simplified profile through the eastern part of the Dales showing the way the Grassington Grit lies on top of various Yoredale beds. The Grit was deposited after parts of the Yoredale sequence had already been eroded away where the Askrigg Block was turned up along the Craven Faults at its southern edge. The profile is 50km (30 miles) long, and the vertical scale is ten times the horizontal scale.

Not only do the faults mark the edge of the Craven Uplands, but they also mark the edge of the limestone country. The faulted boundary between limestone and shale is so clearly seen along the valley east of Gordale Scar, and the walk to the Ingleton waterfalls dives between the limestone walls of Swilla Glen immediately after crossing the South Craven Fault (*see* page 91).

Though difficult to appreciate from surface outcrops, the Craven Faults all dip to the south at 45–60 degrees, and the southern blocks have slipped down over the northern blocks in response to extension of the local crust. This means that they are described as normal faults, but the nature of the faults varies greatly. Where the North Craven Fault is met by the River Twiss above Ingleton (*see* page 91), it is a single clean break, though just a short way down the river (and southeast along the fault) it is a zone of fractured rock 30m (100ft) wide. On a larger scale, the same fault has sub-parallel branches up to 500m (1600ft) apart just east of Settle, but is again a single break past Malham Tarn.

The total vertical movement (known as the fault throw) on the Craven Fault Zone is perhaps best appreciated at Ingleton. The Grassington Grit stands at an altitude of about 700m (2300ft) on the top of Ingleborough. Just south of the faults, and southwest of the village, the old Ingleton coalfield lies in Coal Measures that reach down to about 400m (1300ft) below sea level. Beneath them, the thickness of the Millstone Grit is probably about 300m (1000ft), with the Grassington Grit at its base – about 1400m (4600ft) lower than its position across the faults on Ingleborough. About 200m (600ft) of that difference is the throw on the North Craven Fault, leaving about 1200m (4000ft) throw on the South Craven Fault. Farther east, the North Craven Fault has a throw of about 320m (1050ft) near Malham Tarn, while geophysical surveys indicate about 1800m (6000ft) of throw where the coalesced North and Middle Faults displace the basement rocks.

Faulting though time

The first traces of the Craven Faults were probably fractures developed in the basement under the stresses of the plate collision when the Iapetus Ocean closed about 400 million years ago. The main activity on the faults was right through the Carboniferous, though the fault-bounded Askrigg Block had already risen to a relatively high position before the regional inundation by the Carboniferous seas.

In the early Carboniferous, overall tension caused the Craven Basin to keep dropping so that 2200m (7200ft) of sediments accumulated in it while only 800m (2600ft) was deposited on the Askrigg Block. Much of that difference is accounted for by movement on the Middle Craven Fault. It is known that the faulting was repeatedly active through this time, because so many individual beds thicken markedly to the south as they cross the fault zone. Ground movements were in a complex pattern, and many of the smaller associated faults in the Bordley area, between Malham

The view southeast along the line of the Craven Faults from the limestone benches above Masongill. The dark summit of Ingleborough rises above the white limestone scars of the Craven Uplands, and the slope of the degraded fault-line scarp descends, shadowed by a cloud, to the Craven Lowlands on the far right.

and Wharfedale, were moving, because they too show changes in bed thicknesses across them. In the early Brigantian, part of the fault moved in the opposite direction, down to the north, because the Hawes Limestone thickens abruptly in that direction.

All this movement was taking place within the seabed, and kept the shallow platform of the Askrigg Block apart from the deeper water in the Craven Basin. The steep slope down into deeper water was a prime site for the growth of reefs of coral, algae and all sorts of marine life. These now stand as the Cracoean reef knolls that form the broken line of isolated hills between Ribblesdale and Wharfedale. The edge of the shallow marine platform, marked by most of these, was close to, but not quite along, the line of the Middle Craven Fault. This implies significant fault movements within the Craven Fault Zone at the time, though perhaps on systems of minor faults that are now buried, and are more complex than can be recognized in today's outcrops.

Only a little later within the Carboniferous, the edge of the Askrigg Block moved up on the faults so far that it was above sea level, when hundreds of metres of rocks and sediments were then eroded away. This is clearly recognized, because the Grassington Grit was then deposited right across the eroded ends of so many beds in the area east of Wharfedale.

On a smaller scale, individual fault movements can be recognized where the erosion of steep slopes that were probably newly uplifted fault scarps generated coarse sediment and debris. The Arundian debris flow, just above the basal unconformity at Norber Scar, the conglomerate at Scalebar Bridge and the major local thickening of the Dirt Pot Grit under Grassington Moor, both in the Brigantian (along with the Permian red breccias around Ireby village), all appear to be indicators of repeated movements within the Craven Fault Zone during Carboniferous times.

Each fault movement would have generated an earthquake, but it is very rare for movements during a single earthquake to exceed about 10m (30ft). This is because there is a maximum to the level of stress that can accumulate within rocks before a fault does slip. So the total Craven Fault movements of well over a kilometre must have been accompanied by hundreds of major earthquakes. Many of these were clustered at the beginning and end of the Carboniferous, while the rest were spread through the 30 million intervening years.

Little is known about the history of the Craven Faults since the Carboniferous, because there are no surviving younger rocks that could yield a record of new displacements. It is, however, likely that sporadic movements have continued. It could be tempting to suggest that the very fresh appearance of Giggleswick Scar in particular is evidence of much more recent fault displacements. However, all the big landscape steps along the Craven Faults can be explained by the weak shale outcrops to the south, which have been eroded away so much more rapidly than has the limestone above the fault lines. Where stronger grits lie south of the faults, notably on Bordley Moor and Kirkby Fell, each side of Malham, there is hardly any break in the landscape profile.

Earthquakes are still a result of fault movements, and historical seismic records in the Yorkshire Dales region indicate that there is still some very modest activity. The largest events have been in Wensleydale, in 1768, 1780 and 1933, probably unrelated to any of the major faults. However, the smaller tremor at Skipton, on 30 December 1944, may well have originated from a small slip at depth on the South Craven Fault. It does appear that, though ground movements started along the Craven Faults about 400 million years ago, they may not yet have completely ceased.

The break of terrain along the line of the Middle Craven Fault, seen from the edge of the limestone uplands above Malham Cove. Beyond the limestone slope in the foreground, the small valley drained by Hog House Syke, a tributary of Gordale Beck lies directly along the fault. Its steep rocky slope on the left (north) is in Great Scar Limestone, while the easier soil-covered slope on its right is on Bowland Shales.

levels, with the high blocks known as horsts in between lower levels known as grabens. On land, these create a terrain of basins and uplands. Beneath sea level, they become areas of shallow shelf sea (the horsts) between troughs and basins of much deeper water (the grabens). Such was the origin of the Askrigg Block, rising well above the Craven Basin to its south. The second effect of tension is some overall thinning on the very large scale, and the effect of thinning the crust is to leave its top surface at a lower elevation. As such, the whole region subsided to a lower level, the sea swept in, and the next generation of sediments arrived. These eventually formed the Carboniferous rocks that sit on the eroded ends of the basement rocks at the splendid basal unconformity, perhaps best seen at Thornton Force (*see* page 78).

The Askrigg Block was both tilted and curved down towards the north, where it descends into the Stainmore Trough. From a line that is now roughly under the Wharfedale–Wensleydale watershed, the northern part of the Block was subsiding beneath a sea where more than 1km (1.6 miles) of sediments accumulated, while the southern part was still dry land. These rocks are now buried out of sight, except where they have been lifted to outcrop west of the Dent Fault near Kirkby Stephen. Only later did renewed subsidence lower the whole region slowly below sea level.

Around this time, about 350 million years ago, Britain's piece of continental crust drifted slowly across the Equator. So the marine invasion was by warm seas that were teeming with life. Sediment from nearby landmasses accumulated in the deeper basins, while the Askrigg Block became a shallow submarine platform, on which shell debris formed the almost pure limestones that now characterize the southern Dales landscapes (*see* page 19). The whole area was subsiding to keep pace with the incoming sediments, but the fault blocks were still moving relative to each other – most notably along the line of the Craven Faults (*see* page 54).

The western edge of the Askrigg Block was separated from another uplifted block – the Lake District – by the Dent Fault. This appears to be another major fault, activated along a line of weakness that was established in that earlier continental collision. Its movement has been mainly lateral, as the two massive blocks of rock slipped past each other. At one time the Askrigg Block stood well above adjacent ground of the Lake District, when the Sedbergh conglomerates were formed as debris fans off the Block. The Dent Fault is still a major feature in the topography where it separates the limestones of the Yorkshire Dales from the slates and grits of the Howgill Fells. The latter are formed by an extension of the Lake District rocks that now stand higher than the matching basement rocks within the Askrigg Block, partly due to the lateral shuffling of the tilted blocks during the Carboniferous.

During the Carboniferous, the Askrigg Block was tilting gently down towards the northeast, so the succession of rocks, notably through the Yoredale beds, now thickens in that direction. The tilt ran into the Stainmore Trough, which separates the Askrigg Block from its northerly neighbour, the Alston Block. This was also the direction from which the Carboniferous delta expanded across the Dales region during the later Carboniferous (*see* page 42) to deposit first the sand and mud of the Millstone Grit succession and then the mix of delta-top sediments that became the Coal Measures. This giant delta ultimately filled the great crustal sag that was initiated by the earlier plate extension and had continued throughout the Carboniferous.

Later events

The sag, the delta, the sea and the sedimentation all came to an abrupt end when the whole region stopped being extended and went back into compression as part of the Variscan Earth movements. These were essentially driven by the collision of the Armorican plate, which

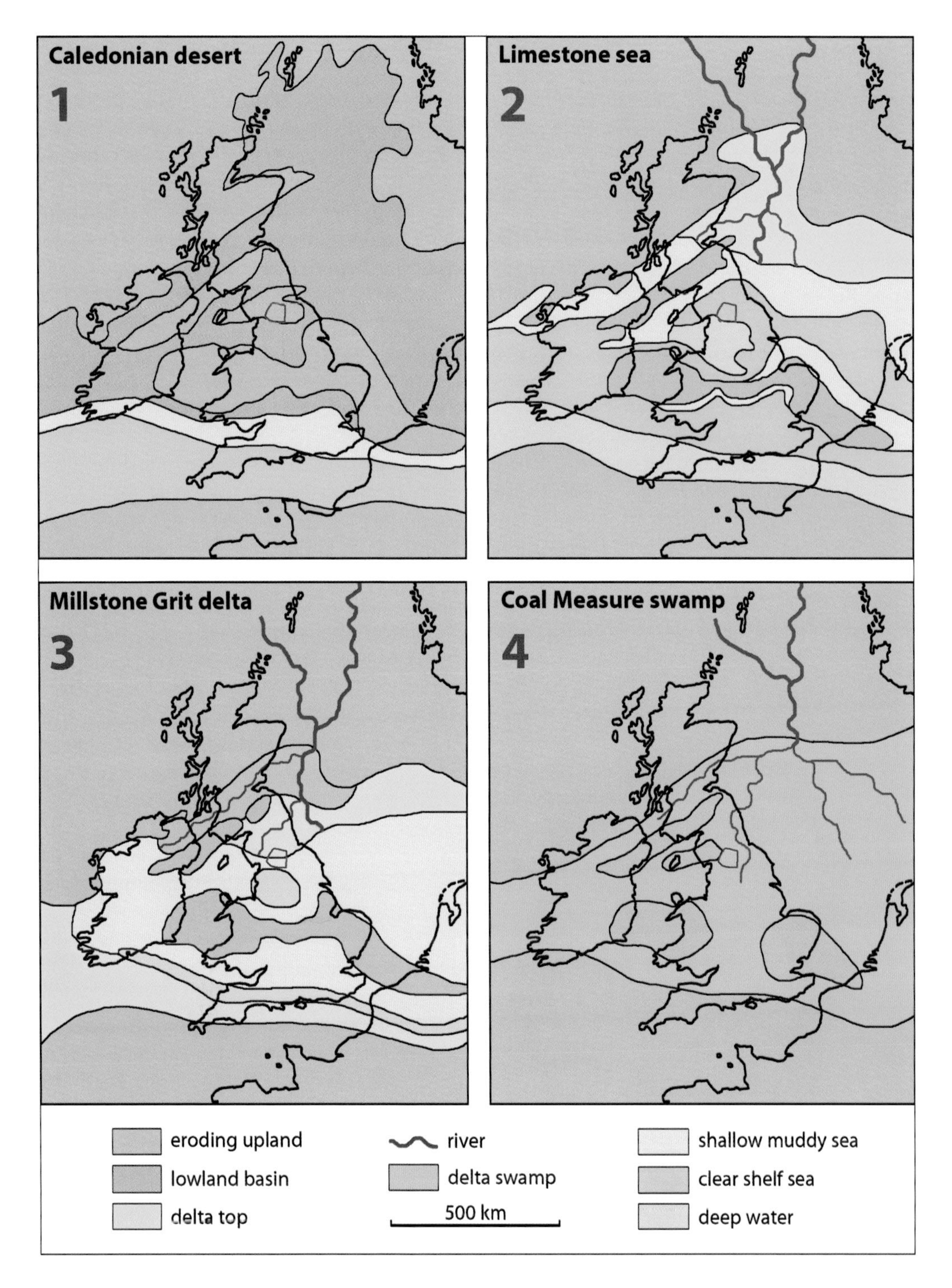

Evolution of the region now occupied by Britain through Carboniferous times, when a new sea was formed and then completely filled. The Askrigg Block, which underlies the Dales, is marked in red. Across southern England, the remnant of deep water in the Rheic Ocean was wider than shown, because its floor has since been crumpled by continental convergence; its southern shore is shown in its correct position beneath France – which was then much further to the south.

Dales Geologist

Adam Sedgwick made his mark in life as one of the great figures in the early years of the science of geology. He was born in 1785, into the family of the Anglican vicar in the village of Dent, so he counts as a true son of the Dales – but most of his work was in the grander theatre of large-scale British geology.

His childhood home life gave him plenty of time to ramble through the countryside, collecting rocks and fossils from the varied sequences that are so well exposed in Dentdale. He went to Sedbergh School, and then onto Trinity College at Cambridge University. In those days there was no formal training in the new sciences of geology and palaeontology, but he stayed on to pursue research and in 1818 he became Woodwardian Professor of Geology at Cambridge. Fieldwork took him to many parts of Britain, and on some of his early explorations in North Wales he was accompanied by a certain young Charles Darwin – providing him with valuable experience in field geology shortly before his journeys to South America and the Galapagos.

Adam Sedgwick (1785–1873).

In 1822, Sedgwick had returned to his northern homeland, though primarily to unravel the complex geology of the Lake District. His fieldwork also took him onto the Carboniferous limestones that form the rim round what he called the Cumbrian slates. At Kirkby Stephen, he described the great downcast fault that separated these limestones from the comparable Great Scar within the Dales – he had recognized the Dent Fault, which passes almost beneath his home and separates the older rocks of the Lake District from the younger rocks of the Yorkshire Dales. And he also recognized the red beds, now known as the Sedbergh Conglomerate, that sit rather out of place between the Lake District slates and the Dales limestone. His papers, read to the Geological Society in London, made giant leaps in unravelling the older parts of England's geology.

Sedgwick was one of several great figures of the Heroic Age of Geology, when the broad framework of geological stratigraphy was established, based almost entirely on fieldwork in Britain. He is perhaps best known for his part in the great controversy of that time. Both he and fellow geologist Roderick Murchison worked on the rocks of Wales, where stratigraphical deductions were hindered by complex folding and faulting of the thick rock sequences. Murchison had unravelled the stratigraphy of a succession of rocks that he named the Silurian System, while Sedgwick was studying an older system that he named the Cambrian. The two geologists presented a joint paper in 1835, on the Silurian and

The splendid chunk of Shap Granite that forms the memorial to Adam Sedgwick in the cobbled main street of his home village of Dent.

Cambrian rocks in England and Wales. Unfortunately, further fieldwork revealed that Sedgwick's Cambrian and Murchison's Silurian overlapped, and each lay claim to that middle slice of beds. They quarrelled and never spoke to each other again. It was not until forty years later that Charles Lapworth resolved the issue by renaming the contentious rocks as the Ordovician System; such is the progress of geological research.

It is perhaps sad that the rocks exposed in the floor of Dentdale, just down from Sedgwick's home village, date from the Silurian that was named by his rival. Adam Sedgwick died in 1873, and is now commemorated by the splendid memorial in the centre of Dent village. The rock that forms the memorial is a chunk of Shap Granite, recognizable by its large pink feldspar crystals, taken from the large quarry on the edge of the Lake District. Geology transcends boundaries, and indeed Sedgwick transcended the Dent Fault for his major work on the older rocks of the Lakes and Wales. But a great geological career, and a lasting component of geology, stemmed from that childhood among the fossiliferous rocks of the Dales.

formed much of central Europe, colliding with the southern side of Avalonia; this closed the intervening Rheic Ocean, which followed Iapetus into oblivion. These Earth movements created mountain chains through southern England and northern France, but northern England was subjected only to modest compression as a marginal effect.

Dominant among the Variscan structures is the Pennine anticline. This is a major upfold that probably developed over an inclined fault, where one basement block rode obliquely up over another, and the covering layers of sedimentary rocks were therefore crumpled into a fold. But the rounded profile of the southern part of the Pennine anticline is lost in the Yorkshire Dales area, where the compression was accommodated by movements on major faults that were already well established. The main effect in the Dales was to raise and tilt the Askrigg Block to roughly its present structure, tilted gently down towards the north and also curving down towards the east. This is best demonstrated by the level of the Grassington Grit, which is about 300m (1000ft) higher on the top of Ingleborough than it is on the far sides of both Swaledale and Wharfedale. Almost the whole of the Great Scar Limestone now dips by just a degree or two towards the north-north-east.

On a local scale, there was some folding of the Carboniferous rocks along the edges of the Askrigg Block, with major faults reactivated in these Earth movements. Close to the line of the Craven Faults, east of Wharfedale, the Greenhow anticline exposes limestone north of the North Craven Fault, while its westward continuation in the Skyreholme anticline brings limestone to outcrop in Troller's Gill, just south of the same fault. Within the Craven Basin, crumpling of beds is slightly more intense in a zone where the rocks were compressed against the southern edge of the Askrigg Block; this is most noticeable in the quarried exposures of the basinal limestones between Skipton and Bolton Abbey. A small syncline contains the Ingleton coalfield, immediately south of the Craven Faults, and this is almost on the line of an earlier anticline – which sadly caused half of the beds containing the coal seams to be uplifted and eroded away even before the end of the Carboniferous (*see* page 48). Against the Dent Fault, the Carboniferous limestones are upturned so

Folded shales and limestones exposed in the old Hambleton Quarry, near Bolton Abbey. These beds lie within the Worston Shales of the Craven Basin, and in this area were crumpled against the southern edge of the Askrigg Block.

Barbondale lies right along the Dent Fault, so the steep slope on the left (west) is formed on strong Lake District greywackes, while the more gentle slope on the right has almost vertical Great Scar Limestone buried beneath a thin layer of glacial till.

that they are vertical in a narrow zone along the flank of Barbondale, from where more modest folding of the limestones extends both north and south along this edge of the Askrigg Block.

It is difficult to recognize post-Variscan movements on the faults, because there are no younger rocks that could have recorded any movements in their structure. There are traces of renewed small movements on the Craven Faults (*see* page 54), and these are likely to have been matched by minor displacements on other faults – further adding to the complexity of the jigsaw of rocks that have been carved into today's Dales terrain.

Following the Variscan uplift, the Pennine areas probably remained as dry land that was being steadily eroded for about 100 million years. A broad and gentle subsidence followed, and it is likely that even the high ground of the Askrigg Block was submerged for various short periods. But any new sediments laid down then have since been completely lost to further erosion, so the history of those times remains conjectural. It is likely that dinosaurs roamed over the low Pennine hills, but they left no trace behind. Since then uplift and renewed subaerial erosion have dominated the Pennine environments. In all, huge thicknesses of Carboniferous rocks have been removed, especially from the axial crest of the Pennine anticline and from the highest of the elevated fault blocks. All the Coal Measures have gone from the structurally high areas, and the Millstone Grit has gone from most of the upturned southern part of the Askrigg Block.

Prolonged erosion has been a key factor in the geological story, because it has exposed the Great Scar Limestone that forms the core of today's Dales landscape. Nearly all that erosion has been by rivers, but the most recent phases were complicated by the periodic advances of glaciers during the Pleistocene Ice Ages (*see* page 68). Rivers and glaciers have deposited the youngest of the sediments in the Yorkshire Dales region; these remain as unconsolidated soils for now, but the future will see their removal too, and they will never be transformed into rock. An upland region like the Dales is not one where new rocks are formed – instead its old rocks are carved into landscapes of character and splendour.

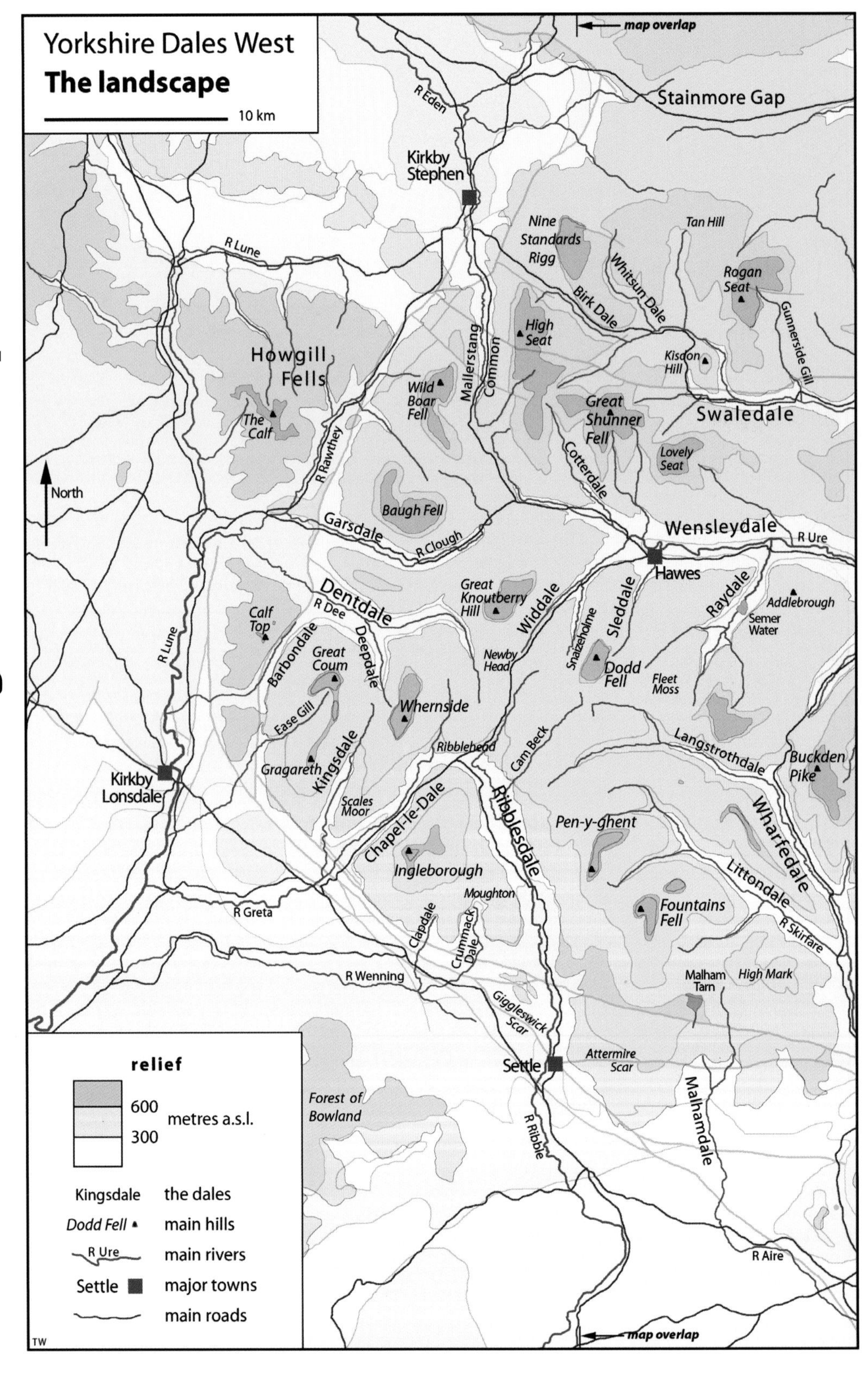

Yorkshire Dales West
The landscape
10 km
map overlap
North
R Eden
Stainmore Gap
Kirkby Stephen
Nine Standards Rigg
Tan Hill
Whitsun Dale
Birk Dale
Rogan Seat
Gunnerside Gill
R Lune
High Seat
Kisdon Hill
Howgill Fells
Mallerstang Common
Wild Boar Fell
Great Shunner Fell
Swaledale
The Calf
R Rawthey
Cotterdale
Lovely Seat
Baugh Fell
Garsdale
R Clough
Wensleydale
R Ure
Hawes
Dentdale
R Dee
Great Knoutberry Hill
Widdale
Sleddale
Raydale
Addlebrough
Semer Water
Calf Top
Barbondale
Deepdale
Great Coum
Newby Head
Snaizeholme
Dodd Fell
Fleet Moss
R Lune
Ease Gill
Whernside
Kingsdale
Gragareth
Ribblehead
Cam Beck
Langstrothdale
Buckden Pike
Kirkby Lonsdale
Scales Moor
Chapel-le-Dale
Ribblesdale
Pen-y-ghent
Wharfedale
Ingleborough
Moughton
Littondale
Fountains Fell
R Greta
Clapdale
Crummack Dale
R Skirfare
R Wenning
Giggleswick Scar
Malham Tarn
High Mark
relief
600
300
metres a.s.l.
Settle
Attermire Scar
Forest of Bowland
Malhamdale
R Ribble
Kingsdale the dales
Dodd Fell main hills
R Ure main rivers
Settle major towns
main roads
R Aire
map overlap
TW

Part II: Creating the Landscape

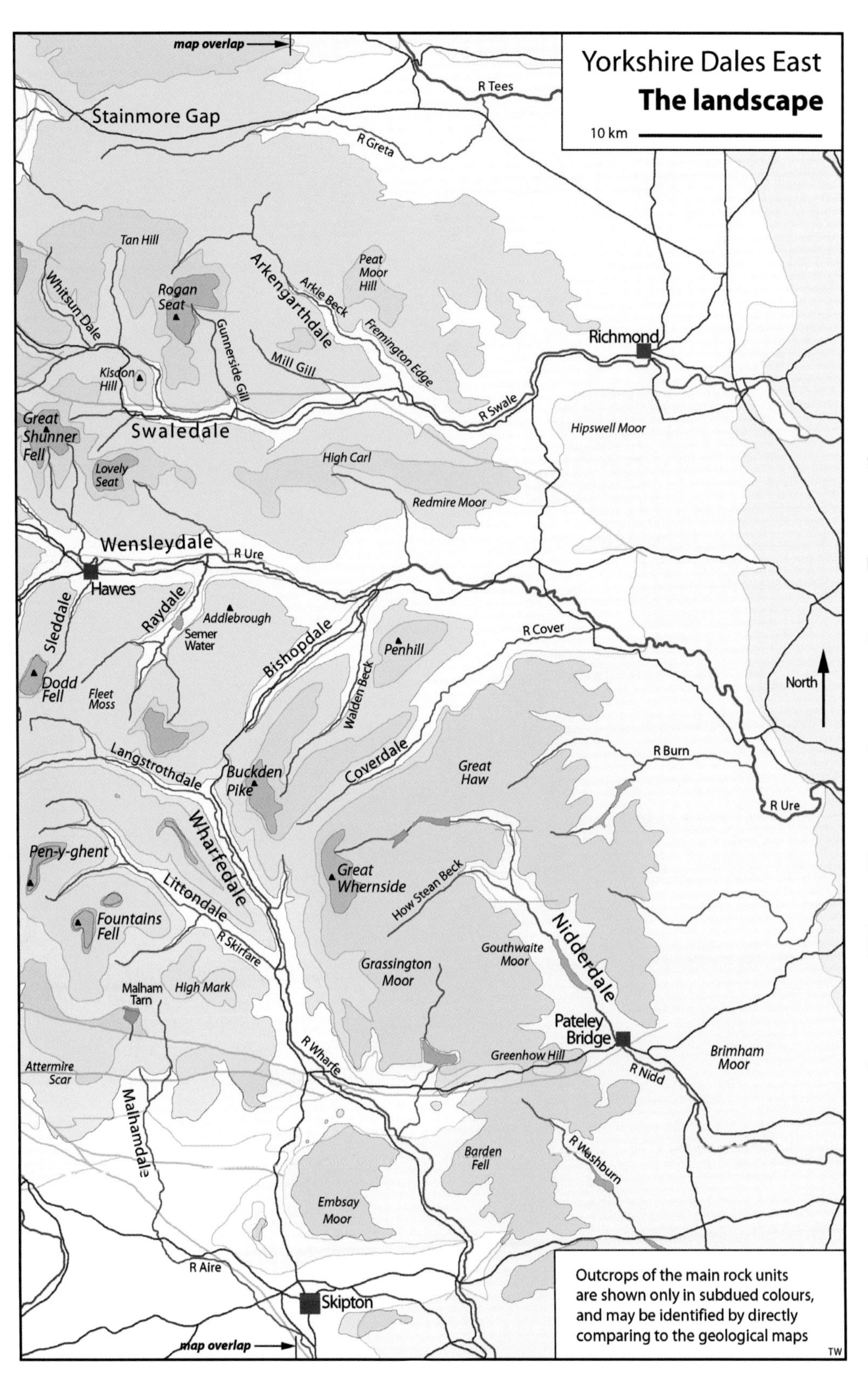

CHAPTER 8

Carving the Dales

Landscapes in the Dales have evolved over some millions of years, and there is no clear start to the whole process. Each phase of erosion removed most trace of earlier events, so today's landscape is dominated by the younger landforms, and rather less is known about their earlier stages. Most details of the Dales landscape date back less than 20,000 years, because they are too small to have survived for much longer. But the major features – the larger Dales and the main summits – would have been recognizable a million years ago. An upland terrain is an environment of erosion, surface lowering, and continual evolution. Agents that fashioned the Dales landscapes have been an interwoven mixture of water and ice, where rivers and glaciers have taken turns to dominate in response to the climatic fluctuations of the Pleistocene. The magnificent glaciated troughs of the Dales themselves are merely the recent features superimposed on an older landscape.

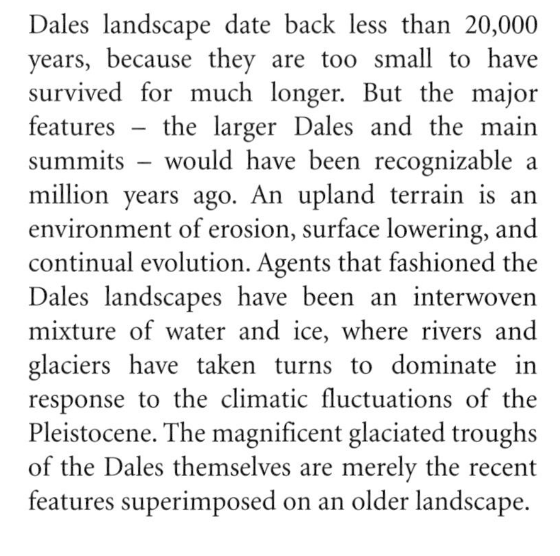

Emerging landscape

Most of northern England, including the Dales area, has been dry land for tens of millions of years. Dinosaurs would have roamed across it 100 million years ago – but then it was a lowland fringe beside the North Sea basin, unrecognizable from today's Pennine hills. Throughout this huge stretch of time, the land was steadily rising, while vast thicknesses of Coal Measures and Millstone Grits were completely eroded away. The erosion was by rivers, and eastward drainage off the uplifted and tilted Askrigg Block saw the ancestors of the Rivers Swale and Ure

A stream is still deepening its own little valley down the side of Coverdale – just one tiny component in the massive and ongoing erosion that creates the Dales landscapes.

Malham Tarn lies on a broad bench that is probably one of the oldest features in the Dales landscapes.

recognizable in roughly their present positions by about five million years ago.

Valley floor erosion is more rapid where gradients are steeper, so steep headwater channels are cut both down and back – in a long-term process known as headward erosion. Drainage of the Craven Uplands towards the south was therefore helped by valleys cutting back into the high ground from the Craven Lowlands, which lay on weaker rocks. In this way the Rivers Ribble and Wharfe became established, though the latter originally drained out past Skipton. This earliest erosion was all by rivers during the warmer climates that preceded the Ice Ages. Given time, rivers erode entire landscapes down to almost level landscapes that lie close to sea level. But they do not achieve this if the land is uplifted – when the rivers have to start again, cutting new valleys to get back down to the new sea level. Sections of flat ground, or matching heights of summits or terraces are recognized as old erosion surfaces that marked these stages in the early evolution of the landscape. The Dales have a lot of ground at a level of about 400m (1300ft), but much is on the top of the Great Scar Limestone, where it is very much guided by the geology. A noticeably level area around Malham Tarn does cut across the North Craven Fault and both limestones and basement rocks; this appears to be a remnant of an old, uplifted erosion surface, but its age cannot be determined.

Once into the last two million years, in the Pleistocene, phases of glaciation could have alternated with the interglacial river erosion, but it is impossible to know how many glaciations there were and what contribution glaciers made to the total erosion of the Dales.

A rough picture of the Ingleborough area just before the onset of the Anglian glaciation can be drawn from evidence that survives in the limestone caves. The floor of Chapel-le-Dale was about 80m (260ft) above its level today, and its adjacent caves in the limestone were flooded below a level of about 280m (920ft). The limestone aquifer had been impounded, and therefore full of water,

Buried by Glaciers

Most features on the Dales landscape have evolved within the last few million years, and this is the time period when the Dales, Britain and the world have undergone a series of major climatic variations. This slice of time is known as the Quaternary, of which all but the last 11,500 years are known as the Pleistocene, or colloquially as the Ice Ages. How much these climatic changes were caused by Earth orbit variations, solar power fluctuations or continental shifts that changed oceanic weather patterns is still open to debate, but there is no doubt that they did happen.

Records from deep-sea sediments show that up to fifty warm and cold climatic oscillations have occurred within the last 2.4 million years. And these are only the major climatic changes, because there were many smaller variations, some only lasting a thousand years, superimposed on the bigger pattern. Major cold phases involved expansion of the continental ice sheets, most notably those centred on Scandinavia and Canada. These phases of outward expansion of the glaciers, across so much of Europe and North America, have become known as the Ice Ages. So the climatic oscillations are known, and have been dated and named, and are also referred to by their Oxygen Isotope Stages as recorded from the marine sediments. But much remains unknown about the exact extent of each phase of glacial cover – because each phase wiped out most of the effects of its predecessor.

The north of England really did have Ice Ages, when it was over-run by massive ice sheets and glaciers that spread out from the north, largely from Scotland and Norway. Maximum reach of the ice cover was roughly along a line from London to Bristol; south of that line, England had only cold phases (with plenty of snow in winter), but there were no glaciers.

So much is unknown about the Ice Ages because the landscapes of today are dominated by features from just two individual Ice Ages – the Anglian, because it was the biggest, and the Devensian, because it was the latest. Anglian ice completely covered the Dales region in its unstoppable slow flow south as far as London and Bristol. There were almost certainly two earlier glaciations of northern England, but no individual landforms and no hard evidence of their passing has survived. There were multiple warm and cold phases between the Anglian and the Devensian, but the extent and effects of the Wolstonian glaciation remain largely unknown (there is even debate over the name of the Wolstonian, but that does not affect the Dales story).

climate	*stages*	*OIS*	*years BP*	*impact on the Dales*
warm	Holocene	1	11,500	post-glacial details
cold	Loch Lomond		13,000	minor
warm	Windermere		15,000	minor
cold	main **Devensian** (Dimlington)	2	24,000	major impact during extensive glaciation
warm/cold	early Devensian	3-4	116,000	uncertain
warm	Ipswichian	5e	128,000	dale incision of uncertain extent
cold	Wolstonian	6	200,000	
warm	Hoxnian	9-11	380,000	
cold	**Anglian**	12	460,000	total ice cover
warm and cold	Cromerian	13		unknown details, broad topography
	earler glaciations	etc		

Significant stages and features of the Pleistocene, along with the Holocene which follows it; together they form the Quaternary. BP = Before Present. These are real ages (actually before AD1950, but the difference of 50 or 60 years is ignorable). The date for the end of the main Devensian glaciation is often given as 13,000 years BP, but that is an uncorrected figure based on raw data from carbon isotope decay (and should be written as 13,000 14C years BP). A true age of about 15,000 years BP (or a date of about 13,000BC) correlates better with data from ice cores and tree rings. OIS = Oxygen Isotope Stages, climatic intervals recognizable in the chemistry of deep-sea sediments.

The Yorkshire Dales during one of the Ice Age glacial advances, with the valleys overflowing with ice while some mountains stand clear as nunataks. This comparison view is actually of the eastern side of the Greenland ice cap.

Also known as the Dimlington stage, or the Last Glaciation, the main Devensian glaciation was hugely important in the Dales, not because it carved more than its share of the overall landscape, but because all its features and landforms are so fresh and dominant today. Glaciers covered all or nearly all the Dales area, and fed into much larger ice streams flowing southwards down the Vale of York and the Irish Sea. Major retreat of the ice started about 17,000 years ago, and left behind a host of glacial retreat landforms that have not been destroyed by any subsequent glaciation. However, the retreat was not uniform, and a climatic blip occurred with another cold phase between about 13,000 and 11,500 years ago. Named the Loch Lomond stage after a large glacier that was then active in Scotland, it produced only cold climates and some tiny glaciers in the Dales.

A major global warming about 11,500 years ago (around 9500BC in calendar terms) marked the transition from the Last Ice Age into the Holocene, which follows the Pleistocene. This is also known as the Post-glacial – or as the latest Interglacial stage, as another Ice Age may be yet to come in future millennia. The last 11,000 years has seen its own climatic oscillations, notably with the warm and wet Atlantic period, around 8500 to 6000 years ago during the Mesolithic (Middle Stone Age), and the Little Ice Age of AD1550–1850, after which global warming has continued until today. These latest events have created very few new landforms, but they have greatly influenced plant growth and have therefore influenced the total landscape. That influence has however been matched by Man, who arrived in the Dales soon after the ice finally retreated, and has created an impact that has steadily increased.

behind the impermeable rocks south of the Craven Faults, until the Dales were cut through that barrier. The Yoredale boundary above the limestone lay close to some large old sinkholes that can still be recognized and dated by their stalagmites (*see* page 145); the limestone was still not exposed around Ribblehead and nor was the basement just above Ingleton. Landscapes in the Dales were comparable with those of today, but better was yet to come.

Anglian glaciation, early and total

This was the big glaciation, when the giant Scandinavian ice sheet coalesced with lesser ice flows from Scotland and the Lake District to sweep south and over-ride the entire Pennines. Across the Dales, the surface of the Anglian ice was about 1200m (4000ft) above sea level, so the entire landscape was buried. While this ice sheet had a top surface that was steadily downhill all the way from Norway to the English Midlands, its base draped itself over the existing terrain. Mountains and valleys were already well established, so much of the ice flow was directed along the main dales, which offered the lines of least resistance; these were the major ice streams where erosion was a little more active along the floor of the ice. Ice flowed sluggishly over the summit of Ingleborough, while it flowed more easily and more quickly down Ribblesdale and Chapel-le-Dale. Faster ice has more erosive power, and so each dale was scoured a little deeper.

Most details of just what the Anglian ice achieved in shaping the Dales landscape have been lost, because everywhere was modified by subsequent erosion. Dales that point towards the source of the ice, which came in from the north and northwest, had ice move up them. Dentdale and Deepdale almost certainly fed Lake District ice up and over into Ribblesdale and Kingsdale (and the same happened again

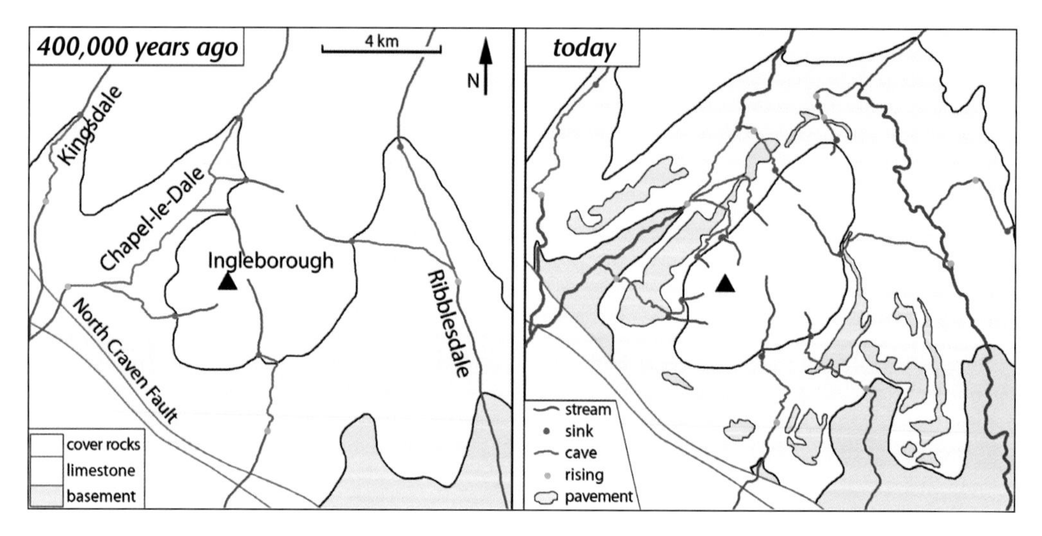

The changing face of the Dales – comparison of the landscapes around Ingleborough nearly 400,000 years ago (soon after the retreat of the Anglian ice cover), with the same area today. Over that period of time, erosion of the shale cover saw the expansion of the limestone plateaus; the extent of the shale cap on Ingleborough is inferred from the positions of stream sinks into caves with dated sediments. Limestone outcrop was reduced where the dale floors were cut down into the basement. The routes of underground drainage, at both stages, are partly known from explored and mapped caves, but are partly conjectured. Limestone pavements were probably extensive after the Anglian glaciation, but they cannot be mapped as all traces were removed by the later glaciations.

The summit of Ingleborough, which was over-run by the Anglian ice sheet. Pavements in the foreground were later trimmed by Devensian ice, and the landslip on the steep Yoredale slope is an even later, post-glacial feature.

in the Devensian glaciation). The head of Snaizeholme is a deep bowl that may well have been scoured out by excess ice in Wensleydale pushing south to escape over the saddle and down into Ribblesdale, but this flow direction was not copied by Devensian ice. Anglian ice might also have flowed south up Bishopdale and over the saddle into Wharfedale. This could account for it being deeper than Wensleydale, to which it is now a tributary, and also for Wharfedale being much larger below the confluence than its source valley of Langstrothdale. It is likely that the original upper reach of the River Nidd flowed east down the valley of the River Burn, before it was captured to flow southwards from the sharp bend above Lofthouse. This was a long time ago as there is no saddle through the modern divide, and the diversion could have been aided by south-flowing ice that crossed the original east–west valley during the Anglian glaciation.

These are only glimpses of what the Anglian ice sheets did to the Dales. Much more remains unseen within the continuous landscape evolution throughout the Pleistocene.

Between the big glaciations

Anglian ice was followed by the great interglacial of the Hoxnian, and then by the short Ipswichian interglacial. In climates considerably warmer than they are today, hippos roamed the Dales, accompanied by elephants, deer, rhinos and hyaenas; bones of all these animals have been found in Victoria Cave beneath a layer of stalagmite dated at 120,000 years old, back in the Ipswichian. Karst processes were active, and many of the stalagmites that still decorate the caves can be dated back to deposition by dripping water that had percolated through plant-rich soils during both interglacials.

Among the various rock shelters along the scars of Langcliffe, Victoria Cave, on the right, has the widest entrance, where sediments inside have yielded evidence of interglacial environments. Much of the grass-covered debris bank below was excavated from inside by Victorian archaeologists, who had found the entrance blocked by sediment and had entered the cave only through the higher fissure on the left.

Erosion by streams and rivers was an ongoing process through these times, but little of its details can be recognized through the later modifications by Devensian ice. An ongoing debate is over just how much the Yorkshire Dales are glacial features. They have clearly been trimmed by glaciers, to give them their splendid U-shape profiles, but they were cut out of the rock by both rivers and glaciers. Even though they are known as the Ice Ages, Pleistocene times had far longer with warm climates than with cold. It is therefore logical that a similarly large proportion of valley excavation was by rivers as opposed to ice. Add to that the pre-Pleistocene episodes of early valley erosion, and rivers should be the main agent.

However, measured erosion rates show that glaciers were important. Rates of floor lowering have been measured for some valleys, where dated stalagmites in adjacent caves indicate when each cave was abandoned because the valley floor had dropped to a lower level. Over the last quarter of a million years, Kingsdale has been lowered by about 60m (200ft), while Ease Gill has been lowered only about 40m (130ft) just on the other side of Gragareth. Kingsdale is a splendid glaciated trough, while Ease Gill is a V-shaped valley that never carried a powerful glacier. The classic Yorkshire Dales were indeed cut largely by rivers, as were most of the world's valleys, but they owe much to Ice Age glaciers that gave them not only their distinctive profiles but also a little extra depth.

Between the two main interglacials, the Wolstonian was a cold stage, but the nature of its erosion processes within the Dales is

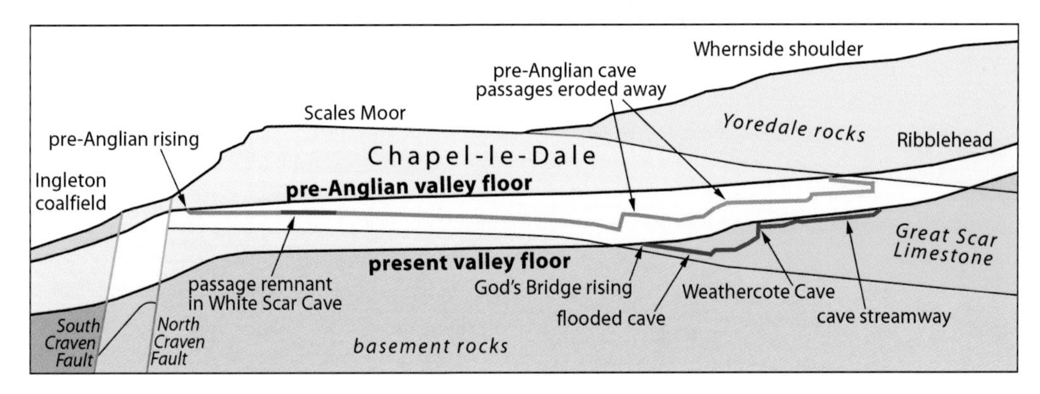

Profiles along Chapel-le-Dale showing the 100m (330ft) of valley-floor lowering in the last half million years, with the pre-Anglian floor profile interpreted from dated cave passages. Most of the old cave passages have been removed by erosion, but the Battlefield Chamber is one remnant that survives in White Scar Cave. The profile is nearly 10km (6 miles) long, with the vertical scale four times the horizontal scale.

unknown. The record of dated stalagmites in the Dales caves does show a gap that was 160,000–145,000 years ago, when none was deposited. This may indicate that solutional activity ceased under a total cover of ice, but the lateral extent of any Wolstonian glaciers within the Dales remains uncertain. Evolution of the Dales did continue through the Wolstonian, but specific landforms have not survived. Glaciers may also have returned to the Dales during the cold stages of the early Devensian, but there was then a long period from 60,000 to 30,000 years ago when a lot of stalagmites were again growing in the caves, and surface conditions were not much different from those of today.

Throughout the Pleistocene, the entire landscape was slowly maturing. The dales were being cut deeper, the fells were weathering just a little more slowly, and the strong limestones were standing out even more on the great scars and plateaux. A mature landscape continues to gain splendour as its local relief is increased, and the Dales landscape was (and still is) a long way off the stage of old age when it is whittled down to flatland.

Devensian glaciation, late and influential

Deteriorating climates during the Devensian caused the Scandinavian ice sheet to expand steadily, until its glaciers reached Britain and tipped it into a major glacial phase less than 30,000 years ago. Maximum ice extent, under the coldest conditions, was about 18,000 years ago, after which there was a steady warming until the ice had all melted away. This then was Britain's Last Glaciation, also known as the Late Devensian or the Dimlington event. Though it lasted little over 12,000 years, it was hugely important to the Dales landscapes. It did not carve the Dales, but it modified almost every aspect of them and provided most of the details. Its impact is so significant because it was so recent that its effects have not yet worn off – which is good, because Devensian landforms include many of the highlights of the Dales.

Devensian ice surrounded and covered the Dales. Scottish ice swept up the Vale of Eden, and was deflected through the Stainmore Gap, then to join more from the north that created the huge glacier down the Vale of York. Lake District ice flowed mainly to the massive glacier south along the Irish Sea, but some scraped round the Howgill Fells and headed down the Lune valley, and some headed east out though the Stainmore Gap.

In the middle of all this, the Dales area was its own ice centre. High ground above the western Dales caught more snowfall from

Landscape and Bedrock

Though weathering and erosion are relentless processes that have steadily lowered the surface across the entire Dales landscape, they are always selective to a degree. Both do respond to the nature and structure of the rocks that form the surface, and they therefore give the geology an undeniable influence in the landscape.

The most obvious geological control in the Dales is the terrain contrast between the three dominant rock types. Limestone creates almost no soil of its own, because it is entirely dissolved into rainwater, so it creates the distinctive karst landscapes with so much bare white rock; the open grassland on limestone is largely on cover soils of hill-wash and transported glacial debris. Grit slowly breaks down to a sandy soil, which clads the grit and supports a plant cover that is dominated by heather across the great moors. Landforms in the mixed sequences of Yoredale rocks are controlled by their weakest link; the shales are so easily weathered and eroded that the whole sequence degrades into gentler slopes. This creates the more open landscapes of Wensleydale, sandwiched between the southern limestone and the northern grit.

Glaciers are notoriously unselective as agents of erosion, as they plough across any terrain. But it is noticeable that the finest of the Yorkshire Dales are carved into the stronger rocks. Chapel-le-Dale and Wharfedale are both steep sided dales flanked by scars of strong, stable Great Scar Limestone, whereas Wensleydale, though fashioned by a more powerful glacier, has had its slopes degraded to more gentle profiles. Within Yoredale country, rock terraces provide a staircase landscape in some parts; some of the limestone scars stood firm beneath the ice and stood proud when the ice retreated, but others have clearly been enhanced by rapid postglacial weathering of the weak shales above and below them.

Finest of all the landforms dictated by geological structure are the great plateaux and slopes that form on top of beds of strong rock. These are known as stratimorphs – surfaces whose shape (morphology) is determined by the rock bedding (strata). Those on the top of the Great Scar Limestone, above the southern Dales are classics of their type. The overlying Yoredale shales have been stripped off, by both passing glaciers and bygone streams, until a thick bed of strong limestone was reached that resisted further erosion. The exposed surface is level in horizontal limestones, but is gently inclined to follow the dip on much of the Great Scar Limestone. A view down Chapel-le-Dale includes magnificent limestone stratimorphs, both level and gently inclined along each side of the glaciated trough. All round Ingleborough, these stratimorphic plateaus are so splendid because much of their limestone surfaces are fretted into great karst pavements of bare, white rock. Almost any Dales

The perfect Dales stratimorph – where a single bed of strong limestone has had its cover of weaker shales and limestones stripped off by Ice Age glaciers; these are the extensive pavements on Southerscales Scars, with Whernside in the distance.

A fine stratimorph, surfaced with its limestone pavements, rises gently to the left following the local dip, while Chapel-le-Dale descends on its right.

pavement is a stratimorph because it is formed on one bed of limestone – including the little one at the top of Malham Cove and the steeply tilted pavements on Hutton Roof Crags. They do all eclipse their cousins on some of the stronger grit beds, which are covered in soil and heather. Patches of stratimorphic surfaces can be recognized on parts of Grassington Moor, but the best is the flat cap on the top of Ingleborough.

Fault scarps are the classic features of geological control in a landscape. But they are formed when a step in the ground is created by the fault movement (which also causes an earthquake); because Britain is now seismically inactive, with few modern earthquakes, there are no true fault scarps in the Dales. But there are fault-line scarps – steps in the landscape created by differential erosion of strong and weak rocks on opposite sides of a fault. The classic example is Giggleswick Scar, on the South Craven Fault just west of Settle. Its long crag of strong Great Scar Limestone overlooks lower ground on weaker shales and sandstones within the Millstone Grit Series. East of Settle, Attermire Scar is a matching scarp along the Middle Craven Fault. On a grander scale, though less spectacular, the whole southern margin of Ingleborough, from Ingleton to Ribblesdale, is a rather rounded fault-line scarp along a zone of multiple faults; the erosional resistance of the Great Scar Limestone creates the boundary between Craven's Uplands and Lowlands, right along the faulted margin of the Askrigg Block. Scattered across the Dales, various lesser crags are fault-line features, but all lack the height and length of Giggleswick and Attermire.

Giggleswick Scar is the textbook example of a fault line scarp, where white crags of Great Scar Limestone overlook weaker shales covered in soil and grass on the other side of the South Craven Fault.

Atlantic winds, and accumulated so much that it became an ice centre. From a broad, gently raised ice dome, glaciers flowing away in all directions. Ice divides did not follow the contemporary or present terrain, as flow is determined by the top surface of the ice, not its base. So ice could move uphill over the rock surface. Deepdale provides the best example of this, where ice swept up it and over the saddle to then flow down Kingsdale; the same ice then continued south to join a massive ice sheet on the Craven Lowlands, which ultimately joined the Irish Sea ice and reached the English Midlands before melting away.

The main ice divides are still recognizable by their minimal glacial erosion where ice was moving very slowly with little driving force. Baugh Fell, Dodd Fell and Great Coum are all rounded and little eroded. The head of Langstrothdale, and its tributary Oughtershaw valley, were little scoured by ice until a glacier developed some power further downstream towards Wharfedale, and the valleys at the head of Swaledale are equally lacking in classic U-shaped profiles. In similar vein, the Howgill Fells, though high and well dissected, lack most of the classic glacial landforms.

While the main Dales ice centre was centred around Baugh Fell during the late Devensian, there were changes in the patterns of ice movement, perhaps driven largely by variations in ice flow from the dominant and adjacent Lake District. Drumlin patterns (*see* page 84) show that ice at Garsdale Head first flowed north to Mallerstang, and later flowed east into Wensleydale. At some time, Dentdale must have had a single glacier right along its deep trough, but this could have been either Lake District ice moving up-dale or Dales ice moving down.

Glaciated dales

The dales themselves are classic glaciated troughs. Straight and deep, with near-perfect

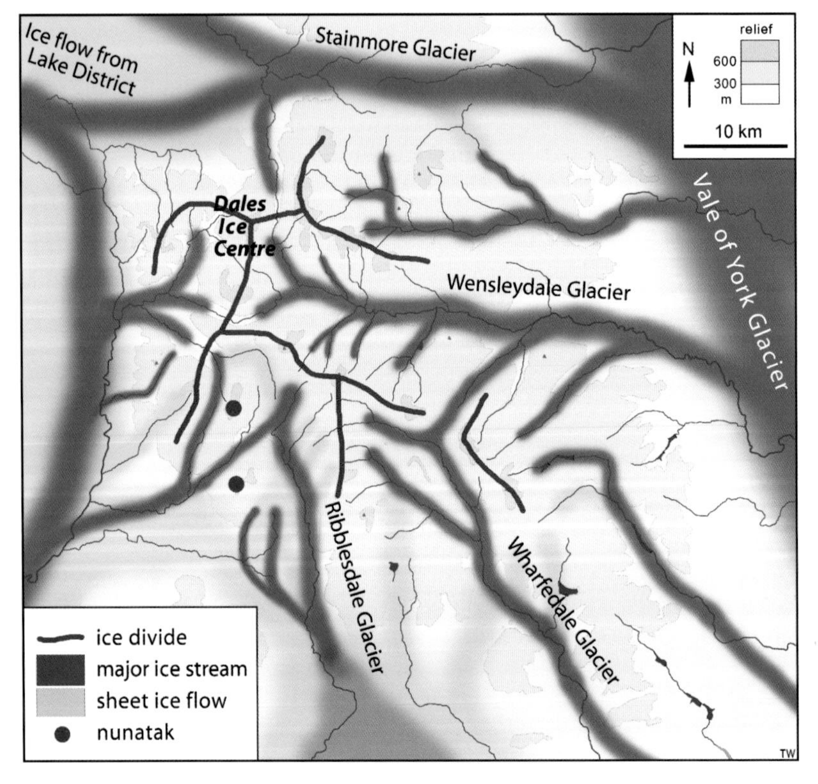

The main pattern of ice flow within the Yorkshire Dales during the Devensian glaciation; note that during the glacial maximum the entire area was covered by ice, except for some peaks along the ice divides and the nunataks shown in red; individual valley glaciers were only apparent towards the end of the glacial stage.

The watershed area east of Ribblehead, which was an ice divide during the Devensian glaciation. In the foreground, the drumlins around Birkwith were left by ice flowing forward and to the right into Ribblesdale. Distant valleys dusted by snow converge at Beckermonds, in the far left of the view, and carried ice away towards Wharfedale.

U-shaped profiles, Upper Wharfedale, Littondale, Chapel-le-Dale and Kingsdale are four of the finest in England, but are only four of the many that characterize the Dales landscapes. It is significant that these four are all cut into the nearly horizontal Great Scar Limestone – because this strong rock survives so well in valley-sides that were over-steepened by ice. Glacier valley profiles are stable when they are cut out, and supported, by the ice; when the glaciers melt away and that support is lost, landslides and rock failures are very common – but not in the strong Dales limestone. Glaciated troughs in the northern Dales are not quite so dramatic, both because the grit is weaker and the varied rock sequences encourage slope failures. The Howgill Fells lack any large glaciated troughs partly because their folded slates do not survive well in the steep upper parts of a U-profile, but also partly because they were at an ice centre with little glacier flow.

Grand glaciated troughs only developed in the dales that carried powerful flows of ice. A major glacier straightened Littondale from a winding river valley into its present magnificent trough, by chopping off older inter-valley ridges to leave truncated spurs. Blue Scar above Arncliffe is one, but the classic truncated spur is Kilnsey Crag just round the corner where it joins Wharfedale. In contrast, and yet adjacent to Littondale, the Malham area was one of slack ice where glacier-carved landforms are notably less impressive (except for the Cove, which had rather complex origins, *see* page 106). An old river valley only evolved into a great dale where ice flowed down it; where the ice flow was across a valley, it retained its old V-shaped profile. Within the eastern Dales, the valleys of Walden Beck, Pen-y-ghent Gill and Cowside Beck all have this profile because ice over-ran them but never flowed down them.

Thornton Force

Panorama at Thornton Force with the modern waterfall on the right and the pre-glacial buried valley next to it on the left – recognizable by the grassy slopes established on the weak glacial till that filled the old valley.

This rather lovely waterfall is popular with all visitors, but it also ranks as one of the finest geological sites in the Yorkshire Dales and perhaps in northern England. It exposes a story of ancient events and then tells another story of much more recent times.

Anyone who sees the waterfall can hardly fail to spot one of its main features, the basal unconformity of the Great Scar Limestone. The lip of the falls is of horizontal white limestone, but the water drops clear to land in a basin cut into vertical Ingletonian slates. The slate was already 180 million years old when it was eroded down to the vertical stumps that formed the Carboniferous lowland of the Askrigg Block. That land was then submerged beneath the sea, where the limestone was formed. It is possible now to walk almost behind Thornton Force and place a hand on that land surface that is nearly 300 million years old. That is the surface that geologists call an unconformity – because the horizontal bedding of the overlying limestone does not conform with the vertical bedding of the slate below. It is so important because it tells of a huge span of geological time – from when muds were deposited, to when they were metamorphosed into slates and folded to vertical, then uplifted and partly eroded away on land, and onward to when the area subsided and the limestone was deposited in a sea that spread over them.

The unconformity can be traced round the wide notch, behind the waterfall, which has been created where two weaker materials have been eroded more rapidly. At the base of the

The classic exposure of the basal unconformity of the Carboniferous behind the falling water of Thornton Force. Above, horizontal Great Scar Limestone. Below, vertical Ingletonian slate. Between them the unconformity, with about a metre of ancient beach bed sediments just above it – with boulders of Ingletonian rock in a matrix of impure limestone.

limestone, a metre of conglomerate reprcsents beach debris created as dry land subsided under an advancing sea. Between large rounded boulders of Ingletonian grit, a weaker matrix of mixed sediment has proved to be softer than the limestone that was subsequently formed in a clear tropical sea. But the rocks on dry land had been well weathered before they were submerged, and their weathered top, a metre or so thick, has also now been eroded out.

Dating back to Ice Age events only thousands of years ago, the second story of Thornton Force is how the waterfall was formed. Pleistocene glaciers flowed down Kingsdale and right on down past Ingleton to merge with the ice sheets on the Craven Lowlands. But as the ice melted away in the face of improving climates about 15,000 years ago, there was a time when the Kingsdale glacier reached not quite as far as the site of Thornton Force – which was not then there. The River Twiss emerged from the snout of the glacier, while the melting ice dumped a huge pile of moraine across the valley – which now forms the ridge of Raven Ray – a textbook example of a terminal or retreat moraine (they are the same, as the moraine was left at the glacier terminus during its retreat). This completely plugged the valley, just to the left of waterfall; where no rock scars break the grassy slope on glacial till, the hillside hides behind it a textbook example of a buried valley.

As the glacier retreated further up Kingsdale, its meltwater was trapped, and created a lake behind thc moraine. A slice of the moraine slumped into the lake, probably soon after the ice retreat left it unsupported. But soon enough, this lake overflowed, not directly

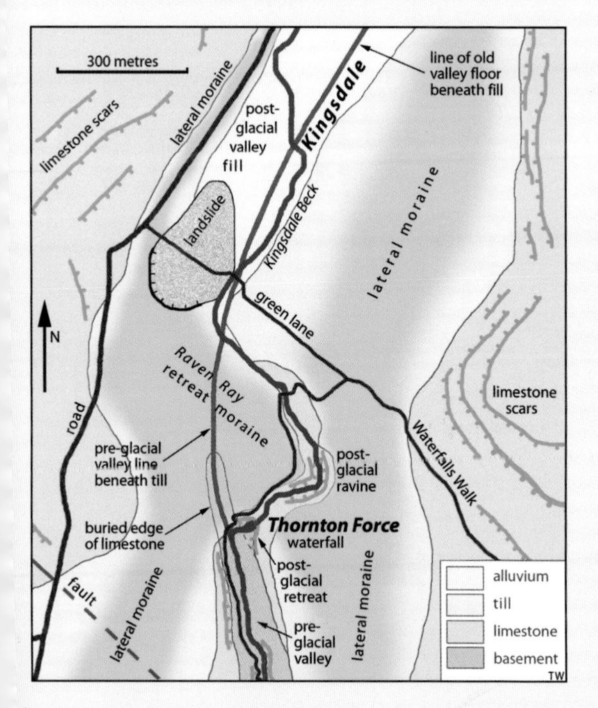

Map of Thornton Force and the buried valley and moraine that led to its development.

1
2
3
4
5

OPPOSITE: **Sequence of reconstructions of the valley at Thornton Force. 1, prior to the Devensian glaciation, the river ran down its valley with a small cascade over the lowest bed of limestone. 2, early in the Devensian, a glacier advanced down the valley (later, this grew so that ice covered the entire scene). 3, at the end of the Devensian, the glacier retreated with water draining from a cave in its moraine-covered snout; after further retreat, the moraine was a complete barrier and a lake accumulated behind it. 4, after the Devensian glaciation had finished, overflow from the lake cut a channel through the moraine with its stream cascading back into the pre-glacial valley. 5, by today the cascade has retreated back to form the vertical waterfall over the base of the limestone.**

above the original course buried under the moraine, but at the lowest point on the moraine crest, which happened to be close to the eastern edge. Water cascaded over the top, and down the old valley side to regain its original course. That tumbling cascade was the ancestor of Thornton Force. Only later did it evolve into a vertical waterfall, when weak slate near valley floor level was scoured out to create a wide plunge pool while the limestone above proved more resistant to erosion. Undercutting by spray erosion behind the waterfall has caused blocks of the limestone to fall away, allowing the entire waterfall to retreat back into its present position. The foot of the original sloping cascade was probably cut back as far as the limestone lip within a few thousand years, and the remainder of its lifetime of perhaps 14,000 years has seen a vertical Thornton Force retreat nearly 50m (160ft) back into the hillside.

Another 14,000 years may see further slow retreat until the waterfall drops into the head of a fine little gorge with limestone walls and a slate floor, along the line of the little ravine already cut since the ice retreat. But the river is already leaking into the limestone, and some of the flow takes an underground short-cut to emerge on the surface of the unconformity to the left of the main waterfall. If limestone fissures are enlarged enough, the entire flow could pass underground; the gorge and the waterfall could become active only during flood events, and eventually even they would cease when an adequate cave matures beneath. Such is the ephemeral nature of a river in limestone country.

The other side of the retreat moraine that buried the old valley – the Raven Ray ridge seen from the limestone crags above Kingsdale. The stream on the right emerges from Keld Head, just below the road.

ABOVE: The classic, U-shaped, glaciated trough that is a Yorkshire Dale, looking up-valley – Upper Wharfedale, seen from the limestone crags above Kettlewell.

BELOW: The classic, U-shaped, glaciated trough that is a Yorkshire Dale, looking down-valley – Chapel-le-Dale, seen from the limestone bench above the Hill Inn, with the profile on the line of the Craven Faults in silhouette against the Craven Lowlands in the hazy distance.

The towering overhang on Kilnsey Crag, on the end of a spur of Great Scar Limestone, neatly truncated by the Wharfedale glacier during the last Ice Age.

Divergence of ice flows also developed. One branch of ice left the Langstrothdale glacier to flow down Bishopdale, and another diverged from the Stainmore ice to develop into a glacier down Arkengarthdale. The glacier in upper Swaledale flowed round both sides of Kisdon Hill; the pre-glacial valley was probably round the west side of Kisdon, but ice breached the saddle over to Swinner Gill, scouring it so low that the Swale still flows that way east of Kisdon. Clearest of all, Ribblesdale ice had a major distributary down Chapel-le-Dale, and then had its west flank spread over the limestone benches of Ingleborough before tightening into distinct ice flows down Crummack Dale and Clapdale. This spread over the limestone benches was a widespread feature, and was the beginning of all the great limestone pavements in the Dales. Ice was not restricted to the deep troughs. For much of Devensian time, these just carried the main flows of ice, while plenty more ice was flowing just a little more slowly across all the high benches and plateaux.

Moraines and erratics

Ice sheets and glaciers dumped till all over the Dales. Till is just glacier debris, which was scraped off the landscape at one place and dropped at another; it is totally unsorted, and is more descriptively known by its older name, boulder clay. A blanket of till some metres thick can provide rather featureless terrain – as around Newby Head, on the saddle between Ribblesdale and Wensleydale, or across so much of the Stainmore Gap. However, an ice sheet slowly moving over such a blanket of till could mould it into a forest of little rounded hills called drumlins. Ribblehead has the finest drumlins, but there are many more around Garsdale Head and down into the Vale of Eden, while the Craven Lowlands and parts of the Lancashire Plain are dotted with forms that are less well shaped.

Away from the Dales ice centre, some of the highest peaks stood above the glaciers to create small nunataks (an Eskimo word for rock peaks sticking up through the ice). They must have done so during the build-up and decline of the glaciation, but whether they did throughout the late Devensian is uncertain. Either way, the main ice flow was deflected around them and slowly moving ice in their lee became zones of till deposition, to form large drift tails. Ingleborough is almost encircled by clean limestone pavements where ice from the north swept over the benches of Great Scar Limestone. Except on its southern side, where till is plastered metres-deep to

Drumlins

Unceremoniously dumped by wasting Pleistocene ice sheets, glacial till can blanket a rock terrain to create a rather featureless chaos of rounded and uninspiring hills in no particular pattern. But in total contrast, till can be moulded into great fields of drumlins that impose slightly ethereal patterns onto the landscape. The Dales has plenty of till, and its drumlin field around Ribblehead is among the finest in Britain.

A drumlin is a small, smoothly rounded, whaleback hill. Typically it is about 20m (60ft) high, 500m (1600ft) long and 200m (600ft) wide, and is slightly steeper at one end so that it has the profile of half an egg (hard-boiled, not fried). Drumlins generally occur in large groups, and are described well by the term basket-of-eggs topography, most similar to where a cloth is draped over the eggs in the basket. Those in the Ribblehead drumlin field cannot be missed in views east from the road south towards Selside. These and all other drumlins are made of almost structureless till, as can be seen in the occasional bank of a stream that has cut into them.

Drumlins are diagnostic features of areas that were once covered by slow-moving ice, generally in broad open valleys. They are aligned with their longest dimension in the direction of ice flow, and the steeper ends point up-flow, though their asymmetry is not always recognizable. Beyond that, it remains unknown just how drumlins were formed. There are plenty of ideas that could be applicable, but there is no single theory that is generally accepted.

It does appear that a zone of ice was already depositing the till at its base, simply because it was overloaded with debris that it had picked up. And this was then shaped into drumlins, because the ice was still moving and riding over its own till. Perhaps each drumlin had a core of frozen till with part-thawed material

The magnificent drumlin field that spreads between Ribblehead and Birkwith.

The farm of Duerley Bottom nestles between two low drumlins on the glaciated floor of Sleddale.

drawn out into the slightly longer tail. Or perhaps each drumlin started to develop beneath a cavity in the ice. There is great debate over the role of subglacial meltwater, which could help shape the drumlins and ease the ice over them, and also whether this was influenced by flood events, when the ice sheets temporarily floated on their own meltwater. The drumlins almost appear like wave forms, but ice moves so slowly that this is difficult to explain; they could be the result of modest pressure waves created within the ice where its flow was deformed between or around existing hills. Too many questions, too few answers.

The patterns of the Ribblehead drumlins do show that they were moulded by ice moving south from Newby Head and were joined by flow from the Cam Fell area, so that they were spread over the limestone scars from Ribblehead to Birkwith and beyond. Also, a few were left by the ice diverging to flow down Chapel-le-Dale. None lies further down either of the valleys where the ice accelerated into the deeper glacial troughs, whereas a few less well-shaped drumlins lie up near the shale boundaries on each side of Ingleborough. This spread of drumlins across the flatter ground at and around Ribblehead does appear to confirm that slow-moving ice was a key factor behind these little rounded hills. Beyond that, drumlins remain one of the more mysterious components of the Dales landscape.

Crina Bottom, seen looking up towards Ingleborough, was the boundary of two regimes of the Devensian ice flow, which headed straight towards the camera. On the left, a powerful ice stream had swept down Chapel-le-Dale and scoured the limestone benches of White Scars. On the right, sluggish ice in the lee of Ingleborough dumped a blanket of till on the limestone, hiding all its scars.

form Newby Moss and the adjacent soil-mantled slopes; in the lee of the summit, it was protected from glacial scour. In similar style, till is conspicuously thick in the upper Darnbrook basin, where it accumulated in the lee of Fountains Fell and its western shoulder that forms Darnbrook Fell. A smaller drift tail extends the southern shoulder of Pen-y-ghent down towards Stainforth.

An even larger drift tail almost fills the Ease Gill basin in the lee of Great Coum. Ice certainly flowed over this summit, but the basin then almost acted as a sediment trap while the sluggish ice crept southwards. Only the post-glacial stream has cut through to bedrock down Ease Gill itself. This basin may hide the longest cave system in Britain, but its surface is a spectacularly unspectacular sweep of moor and bog with just dolines and stream sinks to indicate that a mature karst lies beneath the till.

On a smaller scale, some of the till was dumped at specific locations within the Dales, and so created some distinctive landforms. Kingsdale has textbook moraines – hills of till left behind by the valley glacier. Its retreat moraine – debris dumped at the snout of the glacier when it was fairly stationary for a spell during its slow retreat at the end of the Ice Age – forms the grassy ridge of Raven Ray (*see* page 79). There is also a fine lateral moraine – debris that accumulated along the edge of the glacier, both material scraped off the valley sides and also rocks that had fallen from crags above the glacier; this forms Wackenburgh Hill and the long irregular terrace on which stands the Braida Garth farmhouse. A rather smaller lateral moraine lies on the opposite side of the valley, with the road built along it (though much of this material was originally deposited by meltwater along the glacier margin, so is described by the purists as a kame terrace in part). Prior to forming Raven Ray, a discrete glacier had advanced further, as the lateral moraines extend downstream on both sides until they are lost into the till sheets of the Craven Lowlands. The single narrow ridge of Raven Ray is just one type of retreat moraine, and contrasts with the complex suite of ridges and hummocks across the floor of Wensleydale at Bear Park, just upstream of Aysgarth. Both types can be matched in front of active Alpine glaciers – which neatly represent the appearance of the Dales 16,000 years ago.

Lateral moraines are not so well developed in most other dales, though various fragments may be recognized as elongate ridges and shoulders aligned along valley sides. Wensleydale has conspicuous moraines forming complex ridges and mounds along both sides of the dale around Bainbridge, including the rounded and elongate Brough Hill, on which once stood a Roman fort; this hill has been described as 'drumlinoid' because it was shaped by over-riding ice. Lateral moraines have also been recognized as hillside ridges along the northern flank of the lower dale near Leyburn, but their landforms are not easily distinguished from some of the more modest rock terraces. Wensleydale has various small drift tails, where low, tapering banks of till extend from valley confluences, almost in the style of medial moraines except that the tails were sub-glacial features. Some of these have diverted tributary steams, which now flow along rock benches until they can round the end of the drift tails and drop into the main valley. There they form waterfalls over the harder bands of rock in the Yoredale sequences; Hardraw, Cotter and Mill Gill Forces are all at such sites.

Among glacial deposits within the Dales, special mention of erratics is very appropriate. These are individual large boulders that were carried by glaciers far from their source, and they are especially conspicuous on the bare

Evening sun picks out the green fields that lie on the broad alluvial fan along the floor of Kingsdale. Beyond them, the rounded bank of Raven Ray has coarse brown grass on the glacial till that is the dale's magnificent retreat moraine. Left of the fields, the lateral moraine has slopes of similar texture rising to the dome of Wackenburgh Hill, except where fields have been improved around Braida Garth farm.

limestone pavements. Isolated blocks of grit are scattered across Scales Moor and the pavements on both sides of Chapel-le-Dale. Some could have rolled down from Yoredale outcrops on the higher slopes, but most are too far out, and were clearly dropped by Devensian glaciers. On the other side of Ingleborough, Norber Brow has the most famous of the Dales erratics – hundreds of blocks of greywacke scattered across a limestone shoulder only partly covered in grass and soil. These draw attention, because the greywacke lies beneath the limestone, and so it is often claimed that the erratics were carried uphill – which is proof of transport by ice as opposed to water. In fact they came from outcrops in Crummackdale, and due to a rise in the basement and a gentle fold in the limestone, some of these lie at altitudes above Norber Brow. It is beyond question that the erratics were transported by Devensian ice, but their uphill movement is sadly not provable. On the other side of Ribblesdale, the Winskill Stones are more erratics of basement greywacke sitting on limestone, but uphill transport is not claimed for these.

Glacial retreat

From about 17,000 years ago, the glaciers and ice caps were in steady decay in the face of significant global warming. Within 2000 years, all the ice had disappeared from the Dales. It was only a glacial retreat but it had major impact on the landscape, because features developed at this time were not subsequently over-run, and they still remain very fresh. In these dying years of the Devensian, the last of the glaciers were restricted to flows within the dales. The Chapel-le-Dale glacier would have been a magnificent sight, with a strip of crevassed ice flowing between unbroken walls of limestone not far below the great bare benches; these were barren, snow-swept rock, the pavements of today except that then they lacked most of their fissures, widened by dissolution. On the south flank of Ingleborough, Crummack Dale and Clapdale became recognizable with their discrete glaciers, after most of their excavation had been by localized ice streams beneath an all-covering ice sheet.

The late Devensian climatic improvement was by no means uniform, and the Dales glaciers retreated in stages between slightly cooler periods of relative stability. Sequences of retreat moraines dictate details of the landscape along the floors of the larger eastern

Blocks of greywacke form the train of glacial erratics spread across the limestone shoulder of Norber Scar, on the southern edge of Ingleborough.

RIGHT: Some of the notable landforms left on the retreat of the Devensian glaciers. The corrie glaciers were only developed during the short Loch Lomond stage, after the main retreat.

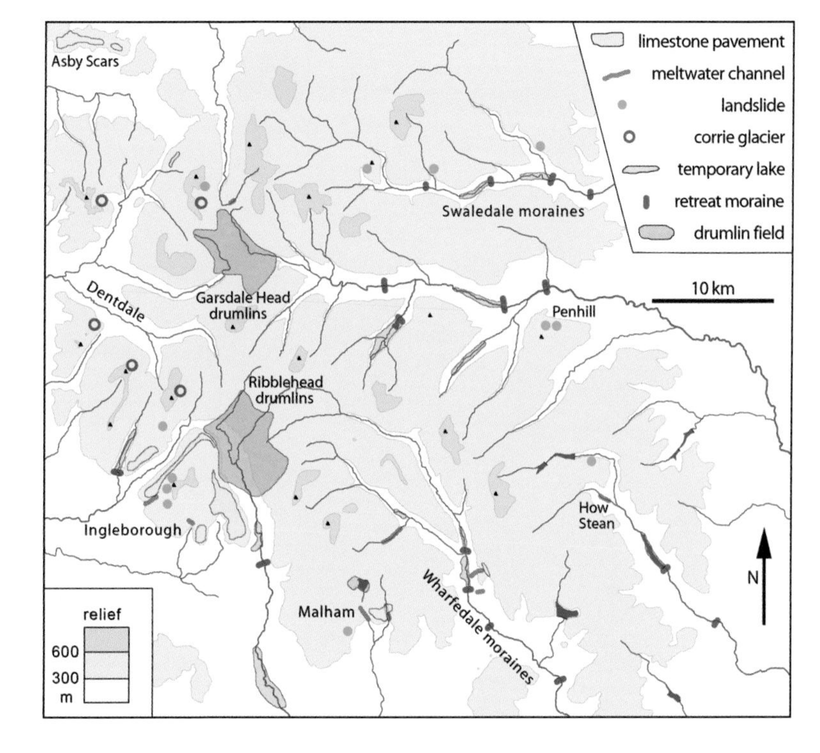

BELOW: Semer Water lies on the floor of Raydale, ponded behind the low remains of a retreat moraine now covered in fields left of the lake.

Walk the Ingleton Waterfalls

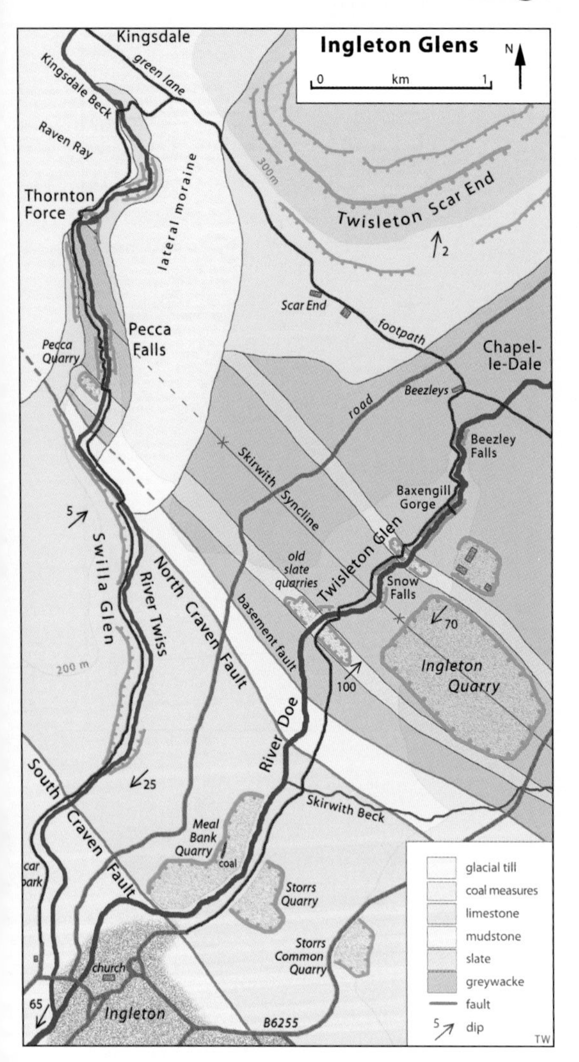

Outline map of the Ingleton Glens; thin soils of till and alluvium cover much of the ground, but only the thick till of the Kingsdale moraines is marked on this map.

A triangular walk round the glens upstream of Ingleton takes in some classic geology – so much so that it is frequently used by groups of students learning the art of geological mapping. There is an entrance fee to the area, which is well justified, because without the made paths and steps, parts of the glens would be rather inaccessible; the two glens now make for a fine half-day's walk. Their names are wonderfully confusing – Swilla Glen lies to the west, and carries the River Twiss which is the downstream continuation of Kingsdale Beck; while to the east, Twisleton Glen carries the River Doe which is the lower part of Chapel Beck.

Head north from the car park towards Swilla Glen. A first bit of wide valley is cut in the Coal Measures that are not exposed, but this ends where the path crosses the South Craven Fault, very close to the small gate. Beyond, the narrow gorge is in dipping Great Scar Limestone, the continuation of which, south of the fault, has been displaced down by more than a kilometre to hide it far beneath the Coal Measures (*see* page 55). A delightful wooded glen takes the path to a footbridge that crosses both the river and the North Craven Fault. The path continues on till-covered mudstones, but the fault is followed by the river, and the far bank is a cliff of bare limestone. The obvious hole in this is an old trial pit that was dug in a failed search for lead ore, on a branch fault that is recognizable between the mudstones and limestone just below. The main fault is lost in trees on the outside of the river bend, where a little scrambling can find it unusually well exposed. North of the fault the limestone is about another 200m (650ft) higher, and has been eroded away within the glen, leaving a little corner of the Askrigg Block's basement rocks exposed along the next section of path.

The mudstones are Upper Ordovician in a thin slice between the North Craven Fault and another minor fault that is lost beneath the soil cover, and separates the mudstones from the slightly older Ingletonian rocks. Just beyond a bank of active tufa on the right (carbonates being deposited by water from limestone that is lost in the trees up the slope), the next bridge across the river leads almost to the mouth of the old Pecca Quarry, which

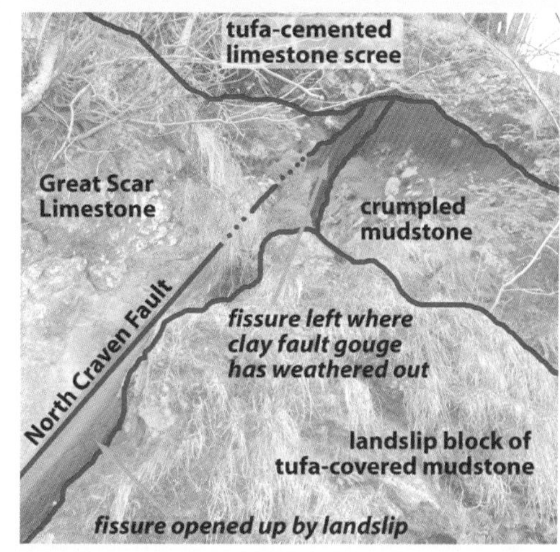

Exposure of the North Craven Fault in Swilla Glen. The dipping fault plane forms the underside of the limestone block on the left, exposed where the mudstones below the fault have slipped away by about half a metre. Across the top of the outcrop, tufa-cemented limestone scree is not broken by the fault because it formed long after movement had ceased. Just beneath this cap, the fault has a zone of clay gouge, formed where the mudstone has been ground to a paste by the fault movement. About 250mm (10in) wide, this has weathered away (partly aided by geology students hacking it out in years gone by), so that it is now barely visible at the back of the fissure.

extracted its not-very-good slate from within the Ingletonian sequence.

The beds here are almost vertical, and immediately upstream, the strong greywackes form the series of cascades that are Pecca Falls. Where the path emerges from the gorge, a refreshment hut marks the axis of the steep synclinal fold in the Ingletonian (though the outcrops that show the change of dip are normally shrouded by the trees). So along the next section of more open valley, the greywackes are repeated and dip steeply southwest, until they are followed by the slates that are seen only in and around the plunge pool of Thornton Force.

Above and upstream of the hut, the crags of horizontal limestone cannot be missed above the path. They wrap round the contours and meet at Thornton Force, where the unconformable base on the slates is so beautifully exposed. But walking towards the Force, spot the grassy bank to the left, which hides the glacial debris plugging the old valley (*see* page 80). The path climbs beside the waterfall and then rises gently out of the post-

Slates stand close to vertical in the old Pecca Quarry.

Thornton Force, where Kingsdale Beck cascades off the limestone and onto the basement rocks for its journey down the glen.

glacial ravine where the river flows over thinly bedded limestone.

Across a third bridge, the path climbs onto Kingsdale's lateral moraine, with the splendid glaciated trough of Kingsdale itself glimpsed to the left, before the route follows the green lane beneath the stepped limestone of Twisleton Scar End.

Beyond Scar End farm, the path eases into the wider and deeper glaciated trough of Chapel-le-Dale – and back onto the Ingletonian basement rocks, though the boundary at the base of the limestone is obscured by glacial till just above the road crossing. The finest of the limestone scars trail into the distance on the left, and the basement greywacke can be seen in Ingleton Quarry to the right. Across the dale, a white house is a landmark for the unconformity, as it stands beside White Scar Cave whose stream emerges right at the base of the limestone. Far above White Scars, Ingleborough rises towards its flat grit summit. Morning shadows might pick out a large ancient landslip in the Yoredale slopes below the lighter band of scars formed by the Main Limestone. Sweeping vistas are a delight on this descent to the River Doe, before the path turns into the woodland shroud of Twisleton Glen.

Beezley Falls are over nearly vertical greywackes, and are best seen where the river steps sharply to the southeast for about 50m (160ft) along a band of weaker slate. From here downstream, the glen varies from wider in the slates to narrower in the greywackes, and both rock types can be seen at footpath level in the appropriate sections. Baxengill Gorge, with a viewing bridge across it, is a narrow cleft along a single joint through fine-grained greywackes, and immediately beyond shallow gullies in the hillsides are the remains of very small old workings that extracted slate from thin bands within the sequence. Steps then descend into a large, old quarry in the Pecca Slate, with a working floor to water's edge, and a matching quarry on the opposite bank. All the slates dip at about 70 degrees downstream, to the southwest.

Even though it is cut through the strongest greywackes – coarse sandstones worked for roadstone in the Ingleton Quarry just above the far bank, the next section of the glen is quite wide because it retains its glaciated profile. Since the glaciers retreated, the river has cut only the trench less than 10m (30ft) deep, below path level, in which lies the modest cascade of Snow Falls. A bridge takes the

The view across to Ingleborough. Above and beyond the terraced crags of Great Scar Limestone, the dark slopes of Yoredale rocks are broken by the paler scar of Main Limestone just below summit cap of Millstone Grit. The white house (standing alone before the large car park was built) marks the base of the limestone where White Scar Cave emerges on its floor of basement slate.

Beezley Falls, where Chapel Beck cascades over greywacke at the head of the glen.

footpath across the river, between walls of almost vertical greywackes – which are the same beds repeated around a very tight syncline; this sharp fold cannot be easily picked out in the wooded glen, but is clearly seen in the quarry above. The path then breaks out into the largest of the old slate quarries, where the slates are slightly overturned, so that their steep dip to the southwest is actually a dip of 100 degrees to the northeast. This is the Pecca Slate again, the same band of good slate that was worked in the quarry above Snow Falls on the opposite side of the syncline, and also in Pecca Quarry, Swilla Glen.

Below the old quarry, the glen stays wide across the poorly exposed Ordovician mudstones, and the footpath then crosses the North Craven Fault unseen. Limestone crags high on the far bank, and then exposed in Skirwith Beck, tell that the basement inlier has been left behind. The path levels out past the old Storrs Quarry, where the geology is best seen in the face of Meal Bank Quarry on the far side of the beck. Massive Gordale Limestones dip at about 25 degrees in this disturbed zone between the two Craven Faults. Prominent across the face is an overhanging bedding plane above a bed of dark grey shale, a metre thick at floor level but thinning up the face. This shale is capped by a thin seam of dirty coal, which formed as something like a mangrove swamp when the limestone seabed was temporarily uplifted. The coal contains pyrite – iron sulphide – which has weathered in seepage water to produce the streaks down the limestone face below; the brown is iron hydroxide (rust) and the yellow is jarosite (an iron sulphate), and not the sulphur that is sometimes suspected.

Dipping limestones in Meal Bank Quarry, with mineral streaks down the face below the exposure of the thin coal seam and shale.

Limestone is lost from sight at the end of the quarries, and the footpath joins a road down through Ingleton. The village straggles across the degraded slope created by the South Craven Fault, descending from the limestone uplands behind to the lowlands on soft Coal Measure rocks ahead. In front of the church, the War Memorial is made of blocks of local greywacke with sawn faces that reveal the bedding graded between coarse and fine layers, though detail is beginning to be obscured by lichens. Down the hill and across both rivers is the car park where the walk started – at the end of a route with few equals in its story of both the rocks and the landscape.

Baxengill Gorge cut along a joint in the strong greywacke.

The very flat floor of Wharfedale, between Kilnsey Crag and Conistone village is on alluvial sediment; this accumulated, shortly after the last glacial retreat, in a lake that was soon lost because it was partly filled and partly drained.

dales, though the sequences cannot be safely correlated. Multiple moraines in these dales prompt the question of whether there were once more moraines in Kingsdale and the other western dales. A late period of cooling may have been enough to cause these glaciers not just to stop, but to advance, wiping out their earlier retreat moraines. Kingsdale, together with Chapel-le-Dale and Ribblesdale, were fed directly from the western end of the Dales' ice centre, where snowfall could have shown different variations because of its direct exposure to wind and weather off the Atlantic.

More conspicuous than the low moraine ridges themselves are the sweeps of flat valley floor on sediments that formed in lakes filled with meltwater between each abandoned moraine and its retreating glacier. Each lake did not survive long, as it was soon either filled with sediment or drained as its overflow cut a new channel through the moraine barrier. Semer Water is the only moraine-dammed lake that has survived, but even this is only a fraction of its former size when it reached 4km (2.5 miles) up Raydale. At times during the ice retreat, the Dales could have been a veritable Yorkshire Lake District (*see* map, page 89) but the shallow lakes were filled or drained that much faster than their deeper cousins in Cumbria, and not all were contemporaneous behind the succession of retreat moraines.

Whereas many patches of level dale floor are signs of bygone lakes, first appearances can be deceptive. The floor of Kingsdale is not as flat as it looks, as its lake flat reaches only halfway from the Raven Ray moraine to Braida Garth farm. Beyond there, the dale floor has a distinct gradient, and is largely on outwash sands and gravels, deposited by braided streams carrying sediment-laden meltwater out of the retreating glacier snout. It was, in effect, an advancing delta, as its lower end reached into the lake, which the finer silts and clays slowly filled in just behind the moraine dam; how long the lake lasted, and how quickly it drained, cannot be known while details of the sediments remain unseen. In Wharfedale, rather similar sediment fans now have the villages of Kettlewell, Starbotton and Buckden standing on them. These are made of debris from tributary streams, which probably formed deltas into shallow lakes within the main dale, though may have been just low alluvial fans onto wetlands on a dale floor not entirely under water. They do have slightly steeper fans onto them, and these were certainly formed above any water level. Arncliffe stands on a very similar fan in Littondale, the expansion of which may have impounded its own lake just upstream, independent of any moraine.

Bishopdale was a little different because the sediments of its flat floor were probably laid down in an ice-dammed marginal lake. This was when the Bishopdale glacier had melted back, probably because it had lost its source when ice ceased to overflow above Cray from a declining Wharfedale glacier, while a reduced Wensleydale glacier did still flow across its mouth. Drainage from Bishopdale was then impounded by an ice barrier, and the lake was formed; till that partially blocks the dale just below Newbiggin appears to be part of a Wensleydale lateral moraine. At a later stage, the Wensleydale glacier left a terminal moraine across its valley just above the site of the Aysgarth High Force, impounding a lake behind it. Overflow water cut channels across the shoulder, and some of this escaped across to drop into Bishopdale, downstream of the site of the lake that was by then probably drained dry.

By about 15,000 years ago, the glaciers had all gone. The warmer climates of the Windermere stage were recorded by new stalagmites growing in many caves, where water percolated down from plant-rich soils mantling the limestone hill. But it was still cool enough to have abundant birch trees in the woodland cover, and more climatic fluctuation was to follow. After just 2000 years, cooling brought in the 1500 years of the Loch Lomond stage. There were no large valley glaciers as far south as the Dales, but little corrie glaciers developed briefly in high basins on Wild Boar Fell, Great Coum and Whernside, as well as in Combe Scar (above Dentdale) and at Cautley Crags in the Howgill Fells. Each of these formed in natural bowls on the northeast side of the hills, which were perfect leeside traps for snow blowing in on the southwest winds and had less than their share of direct solar warmth. These bowls were existing landslide scars, which were further deepened by ice scour, and the landslide debris and glacial till, both of which were left across the front to the bowls, appear today as unsorted soils that are very similar to each other in appearance. Little tarns lay in them once, but are now replaced by small peat bogs – where lay the last glaciers in the Dales.

Meltwater action

Where the Dales' glaciers wasted away, they produced a huge flow of meltwater, and because ice was still occupying many of the main valleys, the meltwater streams eroded valleys and dumped sediments at positions that just do not fit in with the modern drainage patterns. The Black Hill channel, across the Ribble–Aire watershed between Stainforth and Malham, is the classic example, because it is a deeply scoured channel with neither any catchment nor any downstream continuation. It may have been formed by meltwater ponded on the edge of the Ribble glacier that overflowed into the Malham basin, but it could have been sub-glacial drainage beneath a more extensive ice sheet. Marginal overflow channels parallel the valley sides above Wharfedale, and sub-glacial channels are cut into spurs in Swaledale and from Wensleydale into Bishopdale. With the ice now gone, it is difficult to understand those old drainage routes, but their channels are now a part of the landscape.

Above the village of Starbotton, a steep valley down into Wharfedale now carries Cam Gill Beck, but probably owes much of its origins to scouring by meltwater during the glacial retreat. The snow line stands almost on the boundary between the Great Scar Limestone and the overlying Yoredales.

Meltwater erosion has been even more significant to the Dales limestone landscapes, because it flowed on the surface while the ground was still frozen – when caves could not swallow the drainage. Rocky gorges are the hallmark of meltwater on limestone, still there because the strong cliffs survive so well, but dry or nearly dry because the drainage is now underground. Trow Gill, on the flank of Ingleborough, is a splendid meltwater gorge (*see* page 139), but not all meltwater features are such narrow ravines. Draining north from Malham to enter Littondale at Arncliffe, the Cowside valley has a V-shaped profile that shows it was cut by water, but it is far too wide and deep for the very modest catchment of Cowside Beck. Because of its orientation, Pleistocene ice flowed across it and not along it, but wasting ice on the back of Malham Moor would have yielded copious meltwater flows to find the easiest way, under the ice, down to the depths of Littondale.

Malham Tarn lies in a shallow rock basin that was scoured out by ice, but its waters are also held back by a low barrier that lies across its southern flank and also form the little Spiggot Hill that rises from the peat bog beside the tarn. These ridges and mounds are kames – banks of sandy sediment that were laid down by meltwater streams flowing through tunnels along the floor of an almost stationary ice sheet. Their pattern suggests that the sub-glacial drainage may have come from the Black Hill channel and was heading east, probably to drop into a sub-glacial ancestor of Gordale and escape southwards from under the ice. But for most of these meltwater features, the question remains as to how much they were formed beneath the ice or later on in front of the retreating ice.

After the glaciers

Fluctuating and cold climates during the retreat of the glaciers spanned many centuries when much of the ground was frozen, and thawing of frozen ground is prime time for slope failure and landslide activity. Glacially over-steepened slopes in the alternating rock beds of the Yoredales simply became unstable, when weak shales broke down and undermined the strong limestones and sandstones, which then fell in large slabs. Sites of bygone corrie glaciers are not the only landslide scars in the Dales. The slopes around

An evening storm looms behind the series of landslips that score the eastern rim of Wild Boar Fell, overlooking the Mallerstang valley.

Ingleborough's summit have five large landslides, which are easily seen when low evening sun picks out the concave scars with the convex debris piles just below. There is another very similar on Whernside's upper slopes, and Wild Boar Fell is fringed by landslides above Mallerstang. Penhill has two conspicuous slides from Black Scar and Penhill Scar down the slopes into Wensleydale, and Hooker Mill Scar is just the largest of three more around Kisdon Hill in Swaledale.

All these landslides are deep-seated structures, because their shear planes are well down in the bedrock, but they are just part of a series of slope failure types, and grade into shallow soil slides. The latter are particularly prone to fail when the ground beneath is partially frozen, as it was in the Dales at the ends of the glacial stages. Solifluction is the process of downward slumping of saturated soil, commonly but not always over frozen ground, and it is recognizable by lumpy and wrinkled surfaces on steep slopes. Beneath Fremington Edge, the lower slopes of Arkengarthdale are classics of the type. Whether as solifluction or as small landslides, slope failures are extremely widespread on the slopes of Yoredale rocks that are now mantled with soil and debris.

With the ice gone, rivers took over valley erosion. They have trenched though banks of weak soils and glacial till, but their impacts on bedrock have been modest in little more than 10,000 years. Among the most dramatic is the Strid, where the River Wharfe has cut into a

Shadowed by the woodlands of Bolton Abbey, the Strid carries the River Wharfe in a narrow ravine between undercut shoulders of strong Millstone Grit.

At Stainforth Force, the River Ribble cascades over the last beds of strong limestone before crossing a fault onto weaker beds.

strong bed of grit; a series of deep, undercut swirl pools have coalesced along a joint so that the river now flows through their depths in a narrow bulbous slot. The Ure has also cut into bedrock to form the multiple cascades of Aysgarth Falls where it found beds of strong Great Scar Limestone.

Waterfall retreat is always the most obvious sign of river erosion, because the lip of the waterfall is cut back faster than the bed can be cut down, and the result is a short rocky gorge. Hardraw and Thornton Forces (*see* pages 32 and 78) are the classic examples, both where rivers cascade over limestone lips into pre-existing valleys. Wain Wath Force is another, but where the Swale has just met a strong limestone bed while it slowly cuts its valley floor deeper. Stainforth Force owes its origin not to a strong limestone at its lip, but from a fault across the floor of the dale; the original waterfall was created where the Ribble crossed the fault off strong white limestone and onto weaker grey limestone, and that cascade has now retreated nearly 300m (1000ft) upstream, creating a little ravine about 5m (16ft) deep. Linton Falls is similar, where the River Wharfe has cut back into the limestone from an original step onto the weaker ground in the disturbance zone of the North Craven Fault. All these features are post-glacial, though the cascades on the main dale rivers may have been initiated by meltwater beneath the ice.

Post-glacial alluvium, deposited by the rivers, contributed to the early demise of lakes behind retreat moraines, by filling them in as fast as they were drained by down-cutting of their outlets over the barriers of glacial till. Around Helwith Bridge, the floor of Ribblesdale was briefly occupied by a lake dammed behind a bedrock rise capped by a retreat moraine, but it was soon drained to leave the Swarth Moor wetland as the last trace of its short life. Kingsdale has the classic lake site, yet there is no evidence that this survived for long. The main valley floor is a sloping debris apron, and even the flat section immediately behind the moraine is not on lake sediments. It appears that the ravine through to Thornton

Force was cut very quickly to drain the lake – and the lake site has since filled with sediments to an even higher level. The result of this has to been to raise the water level in the Keld Head cave passage, where stalagmites that formed 2500 years ago are now under water.

Peat bogs are entirely post-glacial features. These are formed where dense vegetation holds enough water to keep itself saturated, so that the carbon plant matter cannot be lost to the atmosphere by bacterial action and oxidation. Sphagnum is the classic bog moss, which can form an expanse of shaking ground that looks stable until a hiker walks onto it – and then sinks deep into it. Blanket peat accumulates on high wet moorland, such as on Fountains Fell where some of it is 3m (10ft) thick, and on the great expanse of Fleet Moss astride the watershed between Raydale and Langstrothdale. Fen peat fills in an old pond by sphagnum growth under standing water, so is restricted to discrete hollows; some of these have been artificially drained within the last few hundred years, and Attermire and Linton Mire are just two whose olden names record their wetland origins. Raised bogs form into gentle domes where plant matter continues to grow but retains enough water to survive without loss by oxidation. The Dales' classic raised bog is Tarn Moss, round the western shore of Malham Tarn; its three broad bog domes are formed of peat up to 5m (16ft) thick.

Some of the peat may trace its origins back to nearly 8000 years ago, when the Dales' climate changed into the wetter Atlantic phase. But much of today's bog lands only started to accumulate about 2600 years ago, when the climate became notably wetter and cooler. And by then there was another agent of erosion that was becoming increasingly important in shaping the landscape. Man had moved into the Dales about 12,000 years ago, soon after the main glaciers had retreated. It was Man who cleared the forests, modified the drainage, brought in the sheep and added the walls and villages, which together comprise the last suite of processes in the evolution of the Dales landscape.

Inside the beautiful meandering curves of the River Wharfe near Starbotton, banks of alluvial sediment are progressively colonized by grass, to replace the land eroded away on the outside of the river meanders.

CHAPTER 7

Limestone Country

In terms of its landscapes, limestone is the most distinctive of all rocks – and the Yorkshire Dales display the splendours of limestone on a grand scale.

Limestone landscapes are known as karst, distinguished primarily by their underground features. The Dales are just one style of karst, where the scenery relates not only to the underground streams and rock solubility of limestone, but also to the impacts of glaciation during the Pleistocene, which is therefore known as glaciokarst.

Pavements and karren

Characteristic of glaciokarst are limestone pavements – expanses of smooth bare rock, some flat enough almost to drive a car across, but most cut by networks of deep fissures. The narrow strip of pavement around the lip of

The huge expanse of limestone pavement on Scales Moor, swept clean by the Chapel-le-Dale glacier except that it dropped just a few grit erratics. The benched Yoredale slopes of Whernside rise beyond.

Deep grykes, widened by post-glacial dissolution score the limestone pavements of Fell Close, on the east side of Ingleborough. The Yoredale slopes of Simon Fell rise behind.

Malham Cove is a splendid example, fine enough to be visited by karst geologists from all over the world. Even grander are the pavements of Scales Moor, on the benches west of Chapel-le-Dale, where endless bare limestone creates a moonscape devoid of plant cover and magnificent in its stark beauty.

Good pavements need three factors: ice, limestone and strong rock beds. They are formed wherever retreating glaciers leave a hard, clean rock surface after scouring away all the weak rock, weathered soils and loose debris. They can form on any rock, and are common round the fringes of Greenland and Antarctica. But those left from retreat of the Pleistocene glaciers have been destroyed by more than 10,000 years of weathering that has created a normal soil cover. Except on limestone – where weathering leaves no soil, and the bare rock has therefore survived. In front of retreating Alpine glaciers many ice-scoured surfaces are beautifully clean but are rounded and rolling on massive granite. But where the Pennine glaciers plucked the weaker beds off the Yorkshire hills, they left almost flat pavements on the stronger beds of the Great Scar Limestone – and these survive today in all their glory.

Postglacial rainfall has made its mark on the limestone pavements – by etching out the fissures and grooves that now characterize the pavements, and are collectively known as *karren* (a German word originally). Any fractures on the ice-scraped limestone benches were prone to attack by rainwater, especially if they had been opened up a little by glacial drag. And that seeping rainwater, even though low in biogenic carbon dioxide, attacked the walls of the fractures, so that they are now open fissures, typically 100–200mm (4–8in) wide. These straight, joint-guided fissures are known as *kluftkarren*, and they commonly form networks that frame unjointed blocks of limestone. Locally they are also known as grykes, while the limestone blocks are often

Karst

Karst is defined as a landscape whose features are dependent on the presence of efficient underground drainage. Except in deserts, this is only completely achieved where there are caves large enough to carry streams and rivers, and cave passages are only formed naturally in soluble rocks where the groundwater can dissolve away the walls of narrow fissures to turn them into large caves. So karst is a feature of soluble rocks, of which limestone is by far the most important (but is not the only one). Named after the Kras of Slovenia, karst terrains are found all over the world, and the Yorkshire Dales has one of the finest.

Limestone dissolves in water in the presence of carbon dioxide, which combines with it to make the soluble calcium bicarbonate. The classic school formula simplifies the process (i.e. $CaCO_3 + H_2O + CO_2 = Ca(HCO_3)_2$) but clearly demonstrates the key components. Limestone is the calcium carbonate, rainfall provides the water, and air the carbon dioxide. But as CO_2 is only a small part of the normal atmosphere, the major source of this component is soil air that has been enriched in the gas by the breakdown of dead plant material and oxidation of its carbon. This biogenic carbon dioxide is picked up by rainwater as it percolates through the soil before it reaches the limestone – which it can then dissolve so much more effectively.

Where it seeps down into fractures in the limestone, this aggressive, CO_2-rich soil water progressively develops caves of increasing size (*see* page 123), and a karst landscape can develop. Besides the caves, the obvious signs of a karst landscape are bare rock, dolines, dry valleys and perhaps some distinctive hill shapes.

Bare rock characterizes karst because total dissolution of the rock leaves no mineral residue that can form a soil. Nearly all soils on karst have been transported in – aeolian silt, fluvial alluvium or glacial till. Bare rock survives little changed since it was scraped clean by Pleistocene glaciers, notably as limestone pavements and their intervening scars and cliffs.

Dolines are closed depressions formed where water sinks underground; they are the diagnostic karst landform (and may also be known as sinkholes). Each a metre, or even a kilometre, across, dolines can either be isolated features, or be found in large numbers, which can dominate a karst landscape. Subsidence dolines form within the soil profile, where soil has been washed down into an underlying bedrock fissure; there are thousands in the Dales, all locally known as shakeholes. Dolines in bedrock range between two types named after their processes of formation: solution dolines are essentially valleys with no outlets except underground, whereby the rock has been carried away in solution; while collapse dolines have steep sides left by rock collapse into voids originally created by dissolution.

Profiles through the three types of doline that are common in the Dales.

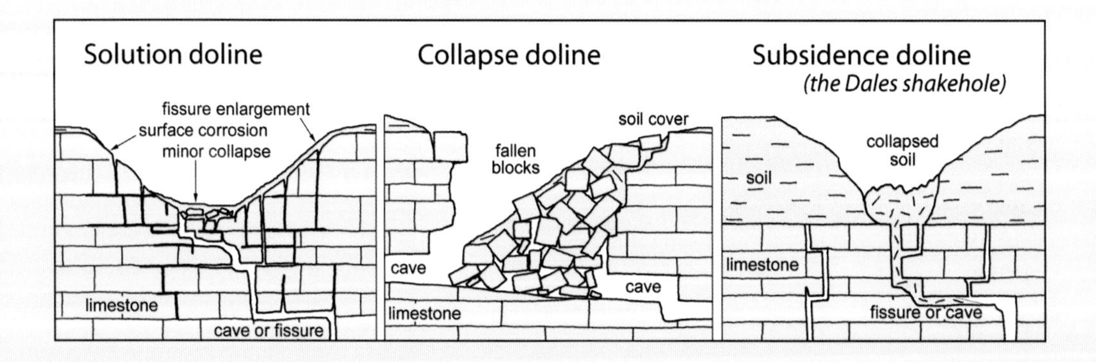

Clints and grykes forming a beautiful pavement on Newbiggin Crags, near Kirkby Lonsdale.

Dry valleys in karst include some that have been carved by streams before the bedrock had caves large enough to swallow the drainage. Others were formed by meltwater, when caves and fissures were blocked by ice in permafrost ground; various periods of such periglacial environment (near to the glaciers but not covered by them) occurred in the Dales during the Pleistocene.

Because limestone dissolution rates relate directly to the amounts of available biogenic carbon dioxide, and these depend almost entirely on the levels of plant activity, karst landforms are significantly influenced by climate. Different types develop in different environments, and this is particularly noticeable on the large scale.

Hot and wet forests in the sub-tropics are perfect for limestone dissolution. Karst in this environment develops giant dolines that pit the entire landscape. These eventually coalesce, so that the terrains evolve to those with only isolated residual hills – including the tall cone hills and towers that are so well-known from paintings of China's karst. These do not develop in drier or colder climates.

Almost no rainfall, means almost no karst in deserts; caves and dolines are commonly left as relics from wetter climates during the Pleistocene. Relict landforms are common in karst, and are often the best evidence of past climates; there are plenty in the Dales that remain from various stages of the Pleistocene.

The third climatic extreme, that of cold and wet, created glaciokarst in latitudes where Pleistocene climates cooled into the Ice Ages. Pavements but not towers, along with dolines and caves, are the karst features that were superimposed on landscapes previously fashioned by glaciers – including those of the Yorkshire Dales.

called clints. Most grykes are only a few metres deep, down a joint through just one bed of limestone and draining into a tiny bedding-plane cave at their base. Grykes wider than about 200mm (8in) were probably partly opened before the ice of the last Ice Age trimmed their crests. The largest grykes are fissure potholes, but they are part of another story (*see* page 125).

Where rainwater drains across the clints, it etches into the limestone by dissolution, and carves out the ever-deepening grooves or rills that are the most common forms of karren. On the Dales pavements, these karren rills may be metres long across a single clint, deepening to perhaps 500mm (18in) before they drain into a deeper *kluftkarren*. They form by deepening of their floors, so their walls are less eroded and tend to have sharp rims – which may just be sharp ridges between adjacent rills; they are then known as *rillenkarren*, and are common on the limestone pavements of the Alps.

However, most Dales karren are much more rounded, in a style that makes them known as *rundkarren*. The rounding is normally developed when they form underneath a soil cover, where the soil and vegetation keep percolation water against all the limestone surfaces. In a few places, notably at the top of Malham Cove, it can be seen that soil has recently been stripped off the pavement along the back margin, so that these *rundkarren* appear to be true sub-soil features. Many areas of pavement once had fairly extensive organic soil cover, created by colonizing mosses, grasses and bushes that have since been lost largely due to sheep grazing. A few protected areas (notably the nature reserve on Ingleborough's Scar Close) have been fenced off to keep out the sheep, so they have re-established a lot of plant cover; these recreate the environment in which

The splendid and often-visited pavement above Malham Cove has deeper grykes in the area emerging from a soil cover.

Landscape features on the southeastern slopes of Ingleborough. Devensian ice scoured the limestone benches to leave extensive pavements around Moughton Scars, before it dropped into the Crummack Dale trough and created the train of erratic boulders across Norber Scars. East of the main flow, sluggish ice in the shadow of the Ingleborough nunatak deposited thick till as it wasted away. The Trow Gill gorge was a later meltwater development. Only the well-developed pavements are marked on this map.

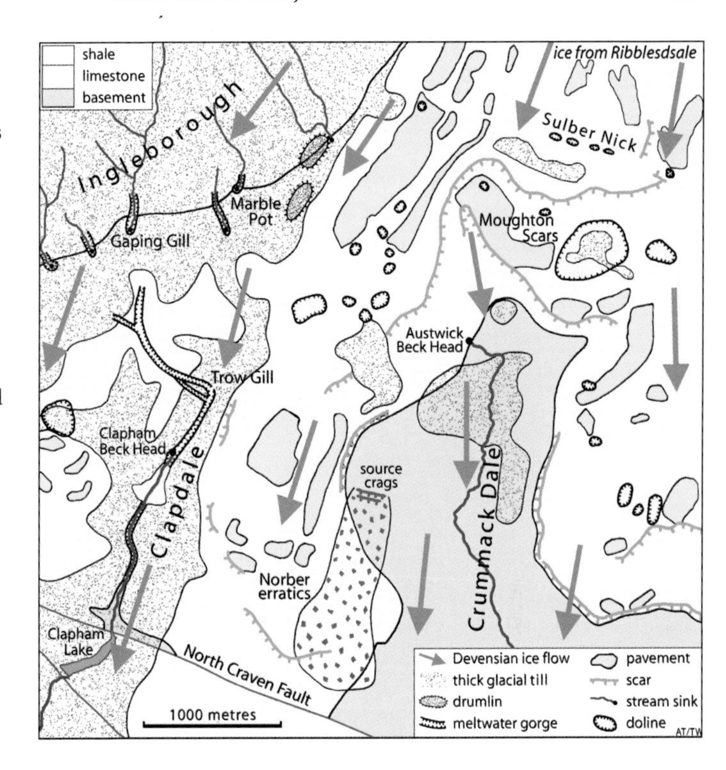

many of the pavements may have evolved. Lichen is a symbiotic assemblage of fungi and algae that creates a thin textured crust on rock, and covers a huge proportion of the limestone outcrops in the Dales. On pavements that never had a rich cover of soil and plants, lichens may have played the critical role of holding the water and dissolving the limestone in order to round off any sharp edges. Thin soils of loess – wind-blown silt that is derived from glacial outwash and the margins of ice sheets – may have once covered some of the pavements; and could then have been lost along with the tree-cover in man's clearances ever since Neolithic times. *Rundkarren* could have developed under the loess, but there is no real sign of its original extent, and there is ongoing debate as to whether loess or lichen was responsible for the rounded edges on the Dales great pavements.

Many of the pavements in the Yorkshire Dales have been formed on just a few strong limestone beds near the top of the Great Scar succession, where thin shales and weaker limestones were easily stripped away by the glaciers and ice sheets. So they lie on the main limestone benches, with a distribution clearly related to patterns of Pleistocene ice movement. The largest areas are on the north and east flanks of Ingleborough that were over-run by ice sweeping down Ribblesdale and Chapel-le-Dale, as well as on Scales Moor, which was also scoured clean by the Chapel-le-Dale glacier. In contrast, there are almost no pavements on Newby Moss and Leck Fell, where great sheets of till were dumped by the glaciers, in the lee of Ingleborough and Gragareth respectively.

Beautiful pavements are scattered across the Dales, with small patches that are well-formed on some of the Yoredale limestones. But some of the finest examples are on the bits of limestone that extend just outside the real Dales area. On the eastern fringe of the Lake District, Great Asby Scar has large areas of magnificent pavement etched by deep *rundkarren*. Towards Morecambe Bay, those isolated limestone hills that were scoured by ice moving south from

Malham Cove

Beloved by acrobatic rock climbers and camera-wielding visitors alike, the Cove's great white wall of limestone, over 70m (230ft) high at its centre, sweeps in a gentle arc to dominate the views north from Malham village. Malham Cove may be a signature feature of the Yorkshire Dales – but it is also one of the most enigmatic landforms in the Dales karst. There is still much that is not known about how it came to be formed.

In simple terms, the Cove is a dry waterfall, where a bygone river dropped over the scarp of the Craven Fault. True, it has been a waterfall. True, the Watlowes dry valley feeds to the top of the cliff. True, it is a feature of the Craven faults, though it has now retreated to a position 600m (2000ft) north of the Middle Craven Fault. But the waterfall theory on its own does not quite fit the picture.

All the water from above the Cove sinks underground into the limestone, and this includes the outflow stream from Malham Tarn, which is lost at Water Sinks. Most re-emerges at Aire Heads, a group of half-choked risings downstream of Malham village, though a small proportion of the flow, from both Water Sinks and other sinks, rises at the foot of the Cove. Water Sinks is a confusing area where the stream sinks through soil and alluvium at various points and in wetter weather maintains a flow progressively further down the valley. Up until about 1850, major storm events caused the Tarn water to flow onwards, over Comb Scar and on down the Watlowes to form a temporary waterfall over Malham Cove. But since then, this has never happened, and the rare flood pulses that have reached Comb Scar have all sunk at its foot. Trickles of water that have flowed over the Cove have all emerged from temporary flood-springs in the Watlowes. Within this period, the sinking water has washed sediment out of the fissures in the limestone beneath and beyond Water Sinks, clearing them out so that they can now swallow even the largest storm floods.

Back in the Pleistocene there were times when an ice sheet filled the wide basin around the Tarn site, but did not reach down to the Cove. Within the largely frozen ground, caves were blocked by ice, so meltwater flowed on the limestone surface – the Watlowes was an active valley, and the Cove was an active waterfall. Furthermore, springtime meltwater floods created very large flows – capable of significant erosion on the face of the waterfall. Perhaps too, some of these floods were hugely enhanced by *jökulhlaups* (that's an Icelandic word for a glacier burst), one of which occurred whenever so much water was trapped under the slowly melting glacier that the ice floated up and let the water escape from underneath. Though *jökulhlaups* only

The great white limestone cliff of Malham Cove.

The dry valley of Watlowes that reaches from Comb Scar down to the head of Malham Cove.

last for a few hours or days, their huge flood flows would have been even more powerfully erosive where they dropped over the Cove waterfall and scoured rock and debris from a churning plunge pool at its foot.

Waterfall erosion alone would have caused Malham Cove to retreat into a narrow gorge not much wider than the river itself – which is clearly not the case. Spring-sapping is another important process, where a cliff is continually undercut by a spring emerging at its foot. Before it lost much of its flow to Aire Heads, Malham Beck had a much stronger flow, and would have efficiently carried away rock debris as the Cove faces retreated by weathering and rockfalls. Beneath the Cove, a cave passage 6m (20ft) wide is now all underwater. For about 150m (500ft) in from the Cove, this was probably an open river passage before the last Ice Age, when its water was ponded behind valley debris that the shrunken stream cannot remove.

The sheer width of Malham Cove suggests that it was cut by more than falling water or a sapping spring – and it is likely that ice contributed to its erosion during the heights of the Ice Ages. The Cove has the proportions of a glacial step, carved out by ice that slowly descended the cliff, plucking away great blocks of the limestone. This was not an individual glacier, but was an ice stream of locally enhanced flow within a wider ice sheet – though appearing as little more than a feature of an ice sheet's floor, an ice stream was a zone of powerful erosion.

So it appears that Malham Cove was knocked roughly into shape far out of sight and far beneath the windswept surface of the Ice Age Pennines. Only later was it trimmed and remodelled by waterfalls, *jökulhlaups*, cave streams and simple weathering. Like so many landforms, Malham Cove is polygenetic – with a host of formative processes active within its complex history.

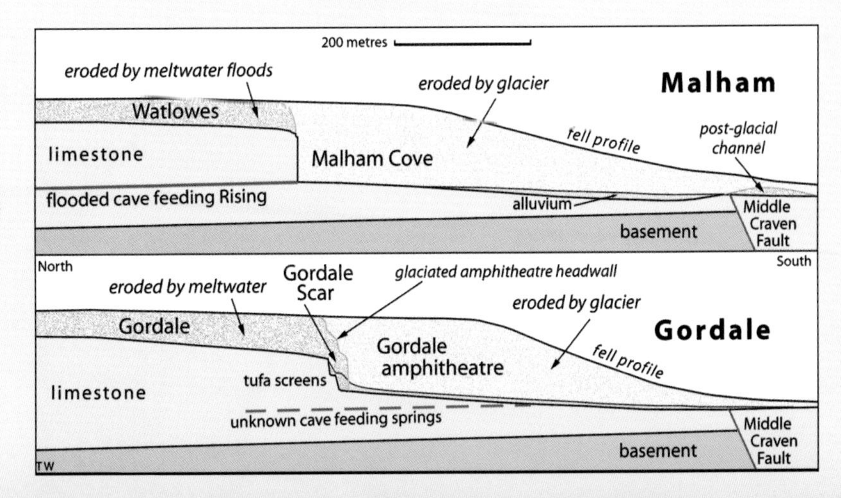

Parallel profiles through Malham Cove and Gordale Scar, showing the similarities in their origins by phases of both ice and meltwater erosion.

ABOVE: The finest of Ingleborough's pavements, on the Southerscales bench above Chapel-le-Dale.

BELOW: Steeply inclined limestone pavements known as The Rakes on Hutton Roof Crag, near Kirkby Lonsdale, with dissolution grooves aligned down-slope between diagonal fractures.

Glacial striae newly exposed from beneath a cover soil of clay-rich till on eastern Ingleborough. Scale is given by the five pence coin. Further away from the soil edge than the coin, the striae have already been lost to dissolution by rainwater.

the Lake District include Hutton Roof Crags, with its spectacular Rakes. These are slabs of limestone dipping at 30 degrees and broken by a grid of oblique, intersecting *kluftkarren*, with each huge, diamond-shaped clint etched by long *rinnenkarren* (each larger than *rundkarren* and with sharper rims) that deepen down the steep slopes.

Ice on the pavements

There is no ice here now, but it was once on all the pavements. Glaciers that drag rocks along their floor leave scratch marks on the bedrock; known as glacial striae, these show the direction of ice movement. But tiny striae, only a millimetre deep, do not survive where exposed rock is corroded by rainwater, so they are not seen on the Dales' many pavements. But striae do survive when protected under watertight clay soil. Over the years, various geologists have exposed striae on a limestone slab near the Long Kin East pothole on eastern Ingleborough, by digging away the edge of the covering clay. Once exposed, the striae are lost to dissolutional erosion within about ten years.

This gives an indication of the short-term rate of lowering of the limestone benches by dissolution, albeit only on the bare rock. A long-term rate may be estimated from the heights of limestone plinths that survive under large, glacial erratic, greywacke boulders. Though isolated small erratics of Yoredale sandstone are scattered over most of the pavements, the most spectacular are the strong greywacke blocks carried from the Dales floors (*see* page 88). Erratics from Crummack Dale were dumped on the limestone bench at Norber, on southern Ingleborough, and some from Ribblesdale form the Winskill Stones on benches below Fountains Fell.

The theory is that limestone under the erratics was protected from rainfall and therefore not dissolved away, creating plinths for the erratics. Traditional measurements at Norber suggest that the plinths are about 450mm (1.5ft) tall, and that implies a dissolutional lowering rate of about 0.03mm/year. This compares to the overall denudation rate of about 0.04mm/year on the limestone, as calculated from the amounts of carbonate carried away in solution from the karst resurgences.

Walk round Gordale and Malham

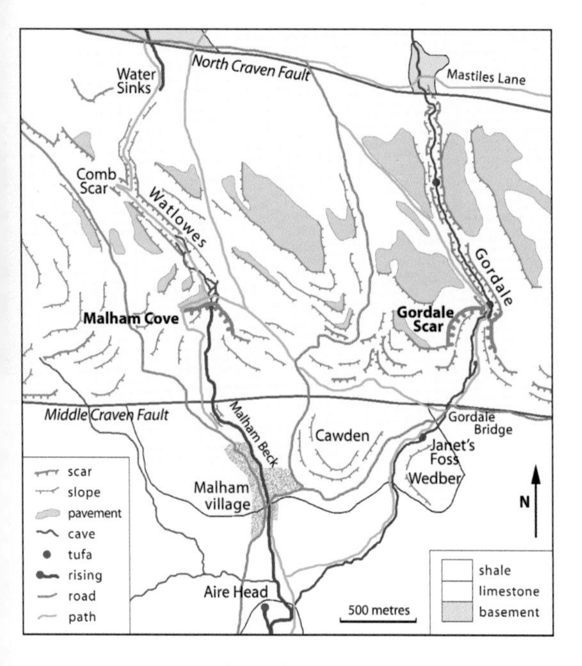

Main features of the landscape around Malham and Gordale.

A circular walk from Malham village takes in both Gordale Scar and Malham Cove, so is popular for its scenic highlights. It also takes in some splendid geology, including the best of the glaciokarst.

From Malham, with its car park, visitor centre and map shop, head south on the Pennine Way. The murky resurgence pools of Aire Head are not worth traipsing across the fields to see. Instead, turn left and follow Gordale Beck up into a wooded gorge that cuts between the hills of Cawden and Wedber, both reef knolls of strong limestone – though the exposed rock is rather featureless. A little further upstream, Janet's Foss is an active tufa cascade 5m (16ft) high, with a splendid screen of moss-covered tufa draped over bedded limestone that dips north off the reef knoll.

The Middle Craven Fault is crossed unseen at Gordale Bridge, from where the path heads north through fields and then passes just above a cluster of springs that pour from the scree slopes. Ahead, the stepped cliffs of limestone rise 150m (500ft) to form a magnificent rock amphitheatre that is directly comparable to Malham Cove, as both were enlarged by Pleistocene ice that poured south off the limestone plateau. However, Gordale's great bowl is broken by a narrower gorge that is almost hidden behind a towering buttress on the approach walk.

Panorama of Gordale Scar with the great amphitheatre of limestone cliffs scored by the meltwater gorge, lost in deep shadow.

Round that buttress, Gordale Scar suddenly comes into full view – a magnificent sight. Far above, the stream falls out of a cave that is actually just the arch of a narrow rock bridge. This eyehole was only formed in 1730, when the water broke through the narrow rib of rock. Before

Janet's Foss, where Gordale Beck slides over a screen of calcite tufa.

then, the stream poured over a waterfall just to the left of the eyehole, its position still marked by the tufa bank that it formed where it landed. This upper tufa, and the notch above it, can only be seen after climbing up another tufa cascade, lower down in the active streambed; this has been eroded into scoops and ledges that offer an easy route in dry weather.

This narrow part of Gordale Scar was formed by an Ice Age meltwater river that followed fault weaknesses until it could cascade into the glaciated amphitheatre. Underground shortcuts through caves were a part of this process until each was unroofed by collapse. One day, far in the future, the rock bridge over today's eyehole will collapse, and thereby contribute to the stream cutting ever deeper into the limestone. Though Gordale Scar is often called a collapsed cavern, this is not the case, as no single large cave ever existed to warrant that description.

Return to Gordale Bridge and take a path that climbs limestone slopes gently westwards towards the top of Malham Cove. Looking back, the line of the Middle Craven Fault is easily seen heading east, between the steep crags of strong limestone on the left and the drift-covered slopes on weak Bowland shales on the right. Across the hill road, the path continues to a fine view of the Watlowes dry valley with the Cove dropping away in front. Now consider how it might have formed by both streams and glaciers (or read page 106). The path traces round the Cove rim and onto that superb stretch of pavement, fretted by beautiful *rundkarren*, polished by innumerable boots and so often photographed.

Steps lead down the west side of the Cove and round to the pool at its foot, where the white limestone cliff rears overhead. There is no sign of the underwater cave that feeds the pool and is still being explored by cave divers. Malham Beck heads down the valley, followed by a lovely path between grassy slopes terraced by ancient lynchets. The Middle Craven Fault is crossed again immediately before the path rises toward the road just before reaching the village. This rock slope has a reverse gradient, in that it was cut by upward-moving ice that came from behind. The beck now runs through a little post-glacial channel down to the left – just one more piece to the story behind Malham Cove.

This circular walk takes a delightful and relaxing half-day. An alternative is to make it a full day by climbing the path up through Gordale Scar, heading north towards Malham Tarn, and looping back on a good path past Water Sinks, round Comb Scar and down Watlowes to the head of the Cove.

The main tufa screens inside Gordale Scar are both washed by waterfalls, the upper one breaking through the hole in a tall limestone rib.

Malham Cove seen from its eastern end, with the Watlowes dry valley emerging over the centre of the cliff.

Glacial erratic of greywacke, on Norber Scar. It stands on a small plinth of limestone that has been protected from some of the rainwater weathering.

However, wind-blown rain and dripwater from the rounded boulders would have reached the limestone plinths and thereby reduced their heights. Furthermore, most of the plinths are clearly higher on one side than the other, and it appears that many of the erratics sit on the edges of very low limestone benches. These mini-benches clearly falsify many measurements of plinth heights, and recent, more careful measurements suggest that truly residual plinths are only about 100–150mm (4–6in) high. This implies a surface lowering rate of less than 0.1mm/year.

That may be a reasonable estimate of the denudation rate for the bare limestone, but the surviving striae indicate a zero rate beneath a cover of watertight clay that covers the rock. In contrast, losses by dissolution at the limestone surface are far greater beneath a permeable soil with a decent plant cover; this is borne out by the higher mean dissolution rate that is calculated from solute loads at the springs.

The other role of ice on the glaciokarst was the trimming of the limestone scars that form the steps between strips of pavements on and down-slope from the main limestone benches. Vertical joints in the strong, horizontal limestone meant that blocks plucked away by passing ice left vertical scars – and few are finer than the multi-terraced Twisleton Scars along the west side of Chapel-le-Dale. Since the ice retreat, frost shattering has broken off chunks of limestone that have fallen to accumulate on the slopes of scree debris; the screes now form long white aprons beneath each scar and overlooking the pavement on the next bench down.

Shakeholes and sinkholes

A closed depression in the ground surface fills with rainwater to become a lake – except in karst, where it drains underground, stays dry, and is known as a doline (or as a sinkhole). There are thousands of dolines in the Dales.

Of the various types of doline (*see* page 102), the most abundant in the Dales are the 'shakeholes' that pockmark all the thin sheets

A cluster of shakeholes breaking the smooth profile over glacial till indicate that there is limestone beneath that forms the floor of Widdale.

of till soils left on the limestone by the glaciers. Rainwater soaks down through the soil and drains into the nearest available fissure in the bedrock beneath. As it does so, it carries tiny particles of soil with it – and begins the process whereby so much soil is washed out that the ground surface collapses and subsides – to form a shakehole (though correctly known as a subsidence doline to geologists). Each shakehole is an inverted cone entirely within the soil, and the limestone may or may not be visible on its floor. Thin soils have lots of small shakeholes, while thick soils have fewer, which can be larger. Thick glacial till dumped in the lee of Gragareth is pitted by some of the Dales' largest shakeholes; Ashtree Hole on Leck Fell is 25m (80ft) across and 10m (30ft) deep, and the soil-choked fissures beneath can be seen in a cave passage 40m (130ft) below the shakehole floor. Most shakeholes mature slowly as their sides steadily slump towards their undermined floors, but larger slumps can sometimes be recognized after a big rain-storm has accelerated downwashing of the soil.

Ashtree Hole, a large doline on Leck Fell has bare slopes of glacial till that are still slumping into choked fissures in the limestone below.

ABOVE: One of the large and very old solution dolines on High Mark, east of Malham, with a person in red by the dry stone wall for scale.

Hull Pot, the large collapse doline on the side of Pen-y-ghent, in moderate flood with a large stream cascading in and disappearing in its boulder floor.

Dolines in bedrock come in various shapes and sizes – the Dales' largest are the saucer-shaped bowls up to a kilometre across on Malham Moor. Though dissolution of the limestone, and its removal by sinking water, are the key processes in forming bedrock dolines, there is usually an element of collapse, where rock is undermined along fissures and bedding planes. Vertical rock walls are the signature of collapse dolines, and Hull Pot, on the slopes of Pen-y-ghent is the perfect example. But that is not to say it is a simple collapsed cavern; Hull Pot probably evolved from a parallel series of wide, fault-guided fissures, where the intervening ribs of rock failed in stages as they were thinned down by ongoing dissolution. In similar style most other bedrock dolines have complex origins involving multiple processes.

True solution dolines are quite rare, though conical funnels a few metres across in polished limestone can be seen at many stream sinks. Variations on the theme are the great shafts and potholes – effectively solution dolines that are deep, narrow and vertical-sided. The Buttertubs, right beside the western hill road from Wensleydale to Swaledale, are splendid potholes with limestone beautifully fluted by water streaming down the walls. The big shafts around Ingleborough are much deeper, but most are less easily viewed.

Some of the larger dolines and potholes in the Dales pre-date the last glaciation and have been partially or completely filled with till. Braithwaite Wife Hole on Ingleborough's northwestern bench is a huge old doline, now with a conical profile created by steep slopes of debris and slumped till that hide the details of its origin. Less obvious are the filled dolines on the benches round the head of Crummack Dale; circles of soil and grass up to 40m (130ft) across are ringed by low rock scars, and give no clue to the depths they reach. In addition, those huge solution dolines on Malham Moor are partly floored with till, so they appear to be preglacial dolines that have been widened out by over-riding ice.

Largest of the Buttertubs, the beautifully fluted potholes in the Main Limestone between Swaledale and Wensleydale.

ABOVE: **The classic stream sink – Fell Beck drops into the deep shaft of Gaping Gill high on the till-covered limestone bench of Ingleborough.**

Dolines can also be called sinkholes, named so because the ground has sunk and water may sink into them. But these are distinguished from stream sinks (or swallow holes). Streams draining from the shale slopes that cap the Dales limestones are lost underground into numerous stream sinks almost along the geological boundary; there are more than a hundred around Ingleborough's summit hill alone. Streams may sink into vertical shafts, into horizontal caves or through chokes of soil and rock debris, each of which may be in the floor of a doline, out on the open benches, or at the end of a blind valley. Mossdale Beck sinks into a choke beneath a low cliff on Grassington Moor, while on Ingleborough, Fell Beck sinks into the deep shaft of Gaping Gill and Alum Pot Beck sinks into the walk-in Long Churn Cave. Each sink is different, but each is just the start of a long cave passage (*see* page 128).

Also on Ingleborough, Marble Pot is a subsidence doline, a stream sink, a pothole, a very large shakehole and a blind valley, all rolled into one. Its great conical entrance is cut into glacial till, which slumps into the open

Marble Pot, on the east side of Ingleborough, after a massive slice of glacial till had slumped into the fissures and chambers in the limestone below.

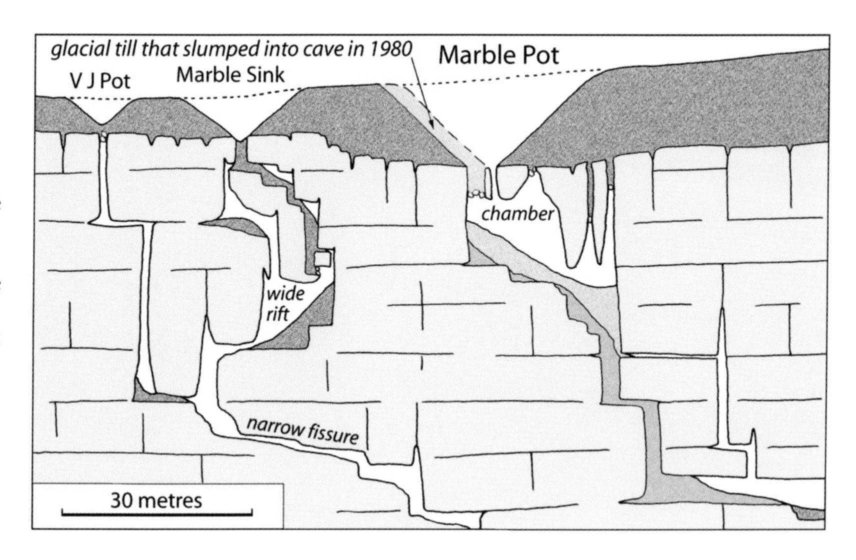

Profile through the trio of large subsidence dolines at Marble Pot, on Ingleborough, showing the slice of glacial till cover that slumped into the shaft in 1980 and added to the debris fill already choking the passages below.

pothole at its foot; effectively it is a giant subsidence doline, which may also be called a very big shakehole. It has a large stream entering on one side and pouring into a cave passage; so it is also a stream sink at the end of a short blind valley, which ends at the shakehole. Inside the cave, the passage is blocked where the water filters through a choke of rock debris; this was dumped in there a few years ago when one side of the giant shakehole above sloughed into the depths. The failed soil slope is still bare, and Marble Pot remains a very spectacular, but rarely visited, feature of the Ingleborough karst.

Dry valleys and gorges

Streams and rivers cut valleys. They do so on limestone – but only where caves are too small to swallow the flow or when they are blocked by permafrost ice. But the valley is then left dry when the stream sinks, either because the stream has shrunk, the caves have matured or the ground ice has melted. The Dales' karst has dry valleys that were formed by combinations of all these processes.

West of Gragareth, Ease Gill was hardly modified by its very sluggish Pleistocene ice and it follows a splendid V-shaped fluvial valley right across the limestone, which is normally dry when the water sinks into the extensive adjacent caves. Partly excavated before the karst had matured, it was modified by sub-glacial and periglacial streams. More open are Conistone Dib and Crina Bottom, each beautiful, sweeping dry valleys. Above its long dry section, Crina Bottom on the west flank of Ingleborough has stream sinks that feed both White Scar and Skirwith Caves, but Conistone Dib is a lovely grass-floored valley, which remains dry except after very heavy rain.

Above Garsdale and just over the watershed into Mallerstang, Hell Gill has cut a splendid gorge across the outcrop of the Yoredale Main Limestone. Its narrow trench, just a few metres wide but 15m (50ft) deep, is just like an unroofed cave passage, with scalloped and polished walls, waterfalls and swirling plunge pools. But it has never had a continuous roof. The surface stream has always cut down faster than caves can be formed along the bedrock joints beneath its channel. Short underground loops formed down joints and out along beddings, but these were only details; some survive as passage fragments in the gorge walls, but many were soon unroofed themselves. If the stream had been smaller, and joints had provided an opportune route, then it would have sunk underground – as have

ABOVE: The lovely dry valley of Conistone Dib, carved by meltwater cascading down into Wharfedale at the end of the last Ice Age.

BELOW: Walls and boulders of the Yoredale Main Limestone are polished and scalloped by the waters of Hell Gill where it cuts down into its gorge above Mallerstang.

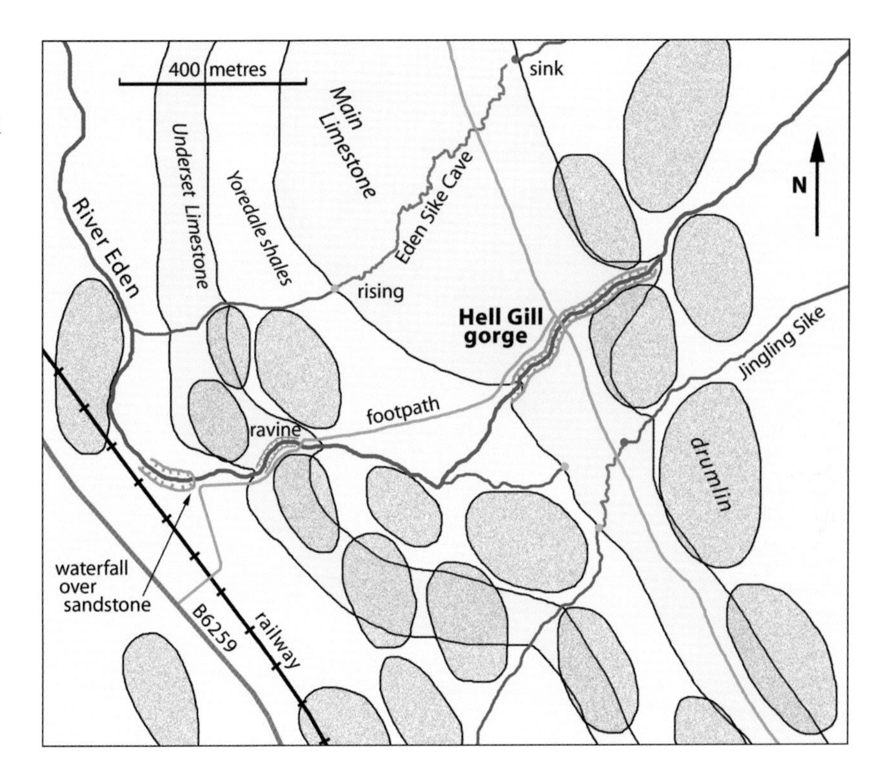

The watershed area on Mallerstang, where Hell Gill is cut into a narrow gorge across the Main Limestone, while adjacent smaller streams pass underground through the same bed, and outcrops of the Yoredale rocks are partly masked by a scatter of drumlins.

most other streams in their journeys across the limestone outcrops.

Meltwater streams emerging from ice sheets over partially frozen ground cut some of the larger new valleys into the Dales limestone. Many of these were scoured out quickly by large flood flows, so they are entrenched into the limestone as some rather fine gorges. Deepest of all is Gordale, just east of Malham. It still carries a small stream, but this is a mere shadow of the powerful river that cut the rocky limestone valley and then plunged into the recesses of Gordale Scar (*see* page 111). Its neighbour, the long, dry valley of Watlowes is also a meltwater feature, whose origins are tied into the story of Malham Cove at its lower end (*see* page 106). Trow Gill, on the flank of Ingleborough, took meltwater that could not sink into a frozen Gaping Gill, and was entrenched by retreat of its cascades down into Clapdale. With its overhanging walls, it has often been thought of as a collapsed cave; but it's not. It was cut by a torrent of meltwater, pouring away from ice sitting on the main limestone benches, and cascading into Clapdale just below its remnant glacier in the bowl of Clapham Bottoms. Waterfall retreat and deep potholes in the swirling riverbed cut the notch back into the limestone, and eating out a few bedding planes to create the undercuts that remain today.

This is not to deny that the cutting and unroofing of little streambed caves always contribute to gorge incision in limestone. The rock bridges that survive for now within Hell Gill, above Mallerstang, demonstrate that process. In Nidderdale, How Stean Gorge is virtually a cave without a roof, originally formed where the limestone structure prevented its stream from finding easier underground routes, perhaps when the ground was partially frozen. These and other miniature gorges are just deep and steep river valleys in limestone. Smaller meltwater features include Conistone Dib and Dibb Scar cut into the eastern flank of Wharfedale. The former has

The great meltwater channel that is the Gordale valley upstream of the gorge through its well-known scar.

a splendid little ravine near its lower end, easily reached because it takes the footpath up from the village; at one point a narrow water-cut slot descends into what was once a deep swirl pool 10m (30ft) across in the meltwater torrent.

Springs and tufas

All the water sinking into the limestone has to return to daylight somewhere, but streams converge underground into dendritic cave systems, so there are far fewer springs (or risings) in the Dales than there are sinks. The classic karst spring is Austwick Beck Head where water that sinks into the various Allotment potholes high on Ingleborough resurges into Crummack Dale; the stream pours from a low, flooded cave on a bedding plane only a metre above the base of the limestone and then cascades out over a rib of slate. Most of the Dales risings are flooded because the water flows south along bedding plane caves that rise on the gentle dip to reach the dale floors. Cave divers have explored many kilometres of flooded passages behind Keld Head, in the floor of Kingsdale, though the water in this dark resurgence pool is also partly dammed back by the sediments in the dale floor. Some cave outlets do flow more towards the north, so they can drain down the northerly dip and out to daylight from open cave passages. White Scar and Dow Caves both do this, at the foot of Ingleborough and Great Whernside, respectively.

The major Dales resurgences carry flows of water that are under-saturated with respect to carbonate, and are therefore capable of more erosion by dissolution. But some smaller springs are fed only by percolation water, which is saturated with carbonate after long and slow journeys through the limestone

The polished slot through a rib of limestone that was cut by a meltwater stream cascading into a wide swirl-pool, near the lower end of Conistone Dib, Wharfedale.

fissures. This water can therefore deposit calcium carbonate when its chemistry is disturbed. The resultant stream deposits are known as tufa, and the best are on Gordale Beck, including the splendid tufa cascade of Janet's Foss, close to Malham village. Certain mosses and algae grow in the lime-rich waters and extract carbon dioxide for photosynthesis, thereby causing precipitation of calcium carbonate to maintain the bicarbonate equilibrium; this happens preferentially where the water is turbulent in its passage over small cascades – which are therefore self-building. Except for its involvement with the mosses, tufa is essentially an outdoor version of stalagmite – just one more variation in the family of karst landforms fashioned out of white limestone.

Austwick Beck Head, with its stream pouring from a cave in the basal beds of the Great Scar Limestone at the upper end of Crummack Dale.

The floor of Littondale, just above Litton village, where the River Skirfare flows underground in normal weather, but floods its limestone bed when the cave passages become full to capacity after heavy rain.

CHAPTER 8

Underground

Beneath the limestone landscapes of the Yorkshire Dales lie about half of all the known caves in Britain. This is because the great plateaux and benches of strong Great Scar Limestone, between the southern Dales, from Malham to Ingleborough and beyond, create the ideal environment to house caves on a grand scale. They lie in the perfect positions below hilltop caps of Yoredale shales that catch the rainfall and pour streams down onto the limestone; Ingleborough is the classic example, with dozens of stream sinks along its shale/limestone boundary. So the streams drop underground, and eventually return to daylight from resurgences at or near the base of the limestone far below in the depths of the dales. Between sinks and resurgences, this natural drainage forms splendid caves – it is doing so today, and has been doing so for over a million years.

Only the climate is not the best for cave development. It is a little too cold in the Pennines (and it was a lot colder during the Pleistocene glaciations), so the plant cover is a bit sparse, and the chemical processes of

Banks of calcite flowstone and a variety of stalactites decorate this corner of an ancient cave passage now abandoned by its stream above Lyle Cavern in Lost John's Hole.

On the western slopes of Pen-y-ghent, Hunt Pot is a classic Dales pothole with a small stream dropping into a deep shaft formed on a joint in the limestone.

limestone dissolution all go a bit slowly. This is why most Dales cave passages are small; walking size maybe, but small compared to the huge cave galleries that lie beneath regions of warmer climate, in places like China or Borneo. Narrow streamways, plummeting waterfall shafts, and a scatter of ancient tunnels decorated with stalactites, these are the main components of the Dales caves.

Shafts and potholes

Inside the nearly horizontal Great Scar Limestone, cave streams most commonly find their way down to lower levels by dropping down a joint – and then developing this into a fine waterfall shaft. Underground shafts are vertical steps within the long underground journeys of the stream caves, while shafts that drop from daylight are the most obvious links between the surface karst and the caves beneath.

Most streams that flow onto the limestone from the Yoredale shale caps, sink down the first available joint. This may drop a stream only a metre or so through a single bed of limestone, to where it flows off along a bedding plane cave. But a major joint or a fault may drop the water much further, through many beds and past many bedding planes. And so are formed the great daylight shafts, often known as potholes, that characterize the Dales karst. Gaping Gill is surely the best known, where Fell Beck drains off the upper slopes of Ingleborough and drops nearly 100m (320ft) down a wide shaft that opens below into the roof of a great cave chamber (*see* page 140).

OPPOSITE: **Major joints and a single bedding plane are picked out in the clean-washed walls of the rift passage midway down Penyghent Pot.**

Across on Pen-y-ghent, Hunt Pot is the archetypal Dales pothole that is easily appreciated from the surface; a small stream drops through a cloud of spray into darkness in an elongate shaft that is clearly a significant joint in the limestone bedrock. Just to its north, Hull Pot offers a complete contrast with its huge quarry-like opening 90m (300ft) long and 20m (60ft) wide and deep. It lies on a minor fault, but there are also many parallel joints that have allowed great slabs of the walls to peel off periodically and crash to the floor, where the rubble is ultimately dissolved by

Caves

Development of a cave passage. Top left, a phreatic tube is formed under water (this one is Boreham Cave, in Littondale). Bottom left, later the water table falls, so that the free-flowing stream cuts a vadose canyon (this one is White Scar Cave, under Ingleborough). Right, a cave that went through both early stages is recognized by its keyhole-shaped cross-section (this is a smaller keyhole passage in the Gaping Gill system, also under Ingleborough).

Caves are an integral part of any karst landscape. Their passages are formed by through-flows of natural drainage. The inflow is either a stream sink, directly into an open cave, or percolation through networks of narrow fissures that only coalesce into a cave passage at depth. The outflow is a resurgence that may be either a freely-draining cave passage or a flooded conduit; this is normally fed by a converging system of caves and fissures; it may be called a rising if it is fed more by percolation water than by sinking streams.

Overall patterns of cave systems are strongly influenced by geological structure. Most passages are initiated along shale beds, on bedding planes, within certain beds, or along joints or faults within the limestone. It may take millions of years to establish an initial drainage route through the network of joints and bedding planes within a limestone mass, but once a through-flow is created, large cave passages can be eroded out of the rock within tens or hundreds of thousands of years. Shapes of the individual cave passages are then largely determined by how they evolved – notably whether they were formed above or below the contemporary water table, and these details are superimposed on the geological controls.

Above the water table, within what is known as the vadose zone, cave passages are free-flowing stream canyons that happen to be underground. In cross section, many are squared canyons with planar roofs along bedding planes or shale beds; others are narrow fissures aligned on vertical or inclined fractures. Long profiles of vadose passages are continually downhill, with meandering canyons between waterfall shafts that are either rounded by spray corrosion or elongated by waterfall retreat; inlets join in a dendritic pattern (with many converging branches) that may include additions through the roof. Vadose caves may be many kilometres long, and depths are only limited by the

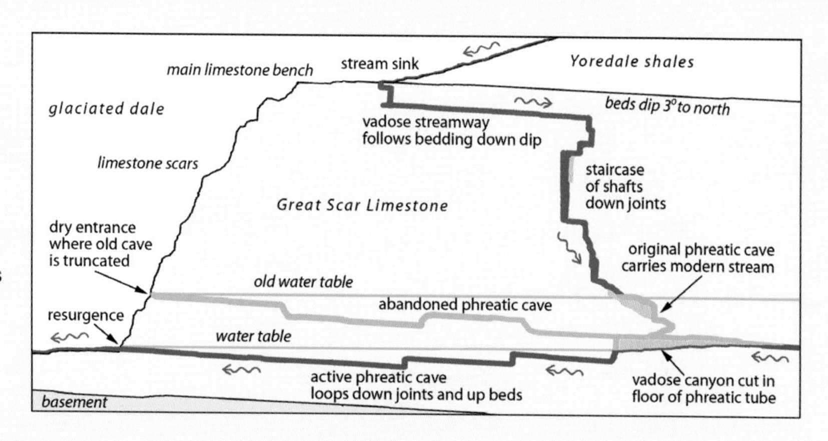

The elements of a typical Dales cave system (loosely based on Swinsto Hole in Kingsdale), with its vadose streamway descending to a flooded phreatic conduit and various high level passages.

vertical range from upland catchment to valley-floor rising. A vadose cave may drain out into a valley, but most continue downstream into flooded phreatic passages.

Below the water table, within what is known as the phreatic zone (or phreas), cave passages are full of water so that dissolution of their walls, floors and roof tends to produce tubular cross sections, either circular or elongated into an ellipse along a joint or bedding. Long profiles of phreatic passages may go up or down, and may be short, flooded down-loops dictated by the geology, or may reach through a long flooded zone in limestone that reaches below their valley-floor resurgence. A phreatic system may be dendritic or braided (dividing and rejoining round multiple loops), or can be a maze, especially if developed on two intersecting systems of joints.

Scattered through many cave system, large chambers are commonly floored by breakdown blocks, and they have clearly been modified by roof collapse. The original formation of these large underground voids is often less easy to interpret, but is usually related to rapid erosion by large cave streams in zones of significant geological weakness and commonly where two or more passages join.

Only in the tropical karsts, where development has not been interrupted by glaciations, do caves mature into large sub-horizontal tunnels graded to their resurgences. In higher latitudes, interruptions to erosional maturity are normal. Notable are the rejuvenations caused by successive Pleistocene glacial deepening of valley floors. These commonly instigate drainage of previously flooded tunnels, so that vadose trenches are cut into the floor of phreatic tubes to create the distinctive cave passages with keyhole cross sections.

Calcite is deposited in the caves in the form of stalactites and stalagmites. Rainwater that has seeped through the soil is enriched in carbon dioxide and therefore dissolves a large amount of calcite from the limestone. But when it reaches an open cave, the excess soil-derived carbon dioxide diffuses into the cave air, and this causes the calcite to be re-precipitated. Old abandoned tunnels are the prime locations for the deposition of dripstone and stalactites on the cave roof and then stalagmites and flowstone on the cave floor. Together with clastic sediments of sand and mud left by declining streams, these form the suite of cave deposits that are especially valuable as datable records of past geological events.

Stalagmite depositions combine with collapse debris from varying degrees of roof failure, and also with in-washed flood deposits, eventually to block many old passages where there is no ongoing stream erosion to keep them open. Alternatively, surface lowering by ongoing erosion (notably by Pleistocene glaciers) can remove the rock and the caves within it – leaving truncated passages behind open entrances in limestone hillsides. Some caves are very old, but nothing lasts forever in an upland environment of active erosion.

rain and stream water. Hull Pot is often dry, because its streams have found new routes down to and through the collapse debris on its floor. But in wet weather a huge stream cascades into it, and after a really heavy storm the stream brings in water faster than it can drain through the debris floor, and the open pothole fills to become a temporary lake. Hull Pot can be called a great pothole or a collapse doline; in reality it is both and is a fine example of how landforms grade into each other.

Hull Pot is huge, but Lancaster Hole is rather less visible, even though it is a pothole more than 30m (100ft) deep. It lies on the grassy slopes above the Ease Gill valley on the western edge of the Dales. Though most of the shaft is over 5m (16ft) wide, it is so narrow at the top that it was originally choked with soil. It was only found by pure chance in 1946, when a passing caver, George Cornes, sat for a rest on an adjacent rock – and noticed the grass blowing in a draught. On a windless day, the current of air could only come from underground, and he soon cleared away enough turf and soil to expose the shaft. He had found the first entrance into the huge Ease Gill Cave System that lies below.

There are dozens of great potholes in the limestone benches of the southern Dales. Whether large or small, they provide an air of mystery where they drop away into depth and darkness, but they are the entrances to cave systems that can reach under entire mountains.

Streamway caves

An underground staircase, with long stretches of clean-washed stream cave punctuated by waterfall shafts, is the typical Dales cave. Water gives life to a cave; it drives the erosion, it creates the activity, it explains why the cave is there.

Between waterfall shafts, many of the Dales' active cave streamways are clean vadose canyons (*see* page 126). The stream erodes the floor, cutting it deeper, while the roof and walls remain almost untouched. So, unlike a surface valley, whose sides flare out by rainwash and tributary erosion, a cave streamway becomes a canyon with vertical walls. Most Dales streamways are canyons cut beneath the bedding planes and shale horizons, still to be seen at roof level, where they were initiated within the limestone. Some canyon caves are only a metre deep below their bedding roofs; others are tightly meandering slots up to 20m (60ft) deep. The width of a canyon is the

The wide canyon passage known as Hensler's Master Cave in the middle reaches of the Gaping Gill cave system.

A tall and narrow canyon passage, with ledges formed along less soluble beds of limestone, in the far reaches of White Scar Cave.

width of its stream, perhaps less than a metre, but 3–5m (10–15ft) in the great master caves that gather inlet streams to drain entire plateaux.

Among the finest of the Dales cave streamways, and certainly the most accessible, is in the Long Churn Cave that drains towards Alum Pot under the flanks of Ingleborough (*see* page 135). Just a few metres high and wide, its stream drains across a polished floor over cascades and through pools between walls that are a myriad of tiny scallops cut into the lovely pale limestone. It is the classic vadose canyon, with wide notches at roof level marking the bedding planes on which it was initiated; its roof steps down a few beddings, and the canyon even has some dry loops where older courses have been abandoned while the stream cut deeper.

Because vadose canyons are cut by free-flowing streams and also follow the bedding, they normally follow a direction down the gentle dip that affects most of the Dales limestones, so these stream caves drain mainly to the north. Most of the major caves are in the thick Great Scar Limestone along the southern edge of the Dales – where the ground surface and the main valleys drain to the south. Consequently, most of the large cave systems extend downwards into passages that still follow the bedding but have to carry the drainage back up the dip towards the resurgences in the lowest parts of the Dales floors. So these lower passages are mainly flooded – forming the phreatic tube caves that are described below. The only long vadose streamway to follow the bedding down-dip, out to its resurgence at the surface is White Scar Cave, draining northwest to emerge in the southern flank of Chapel-le-Dale.

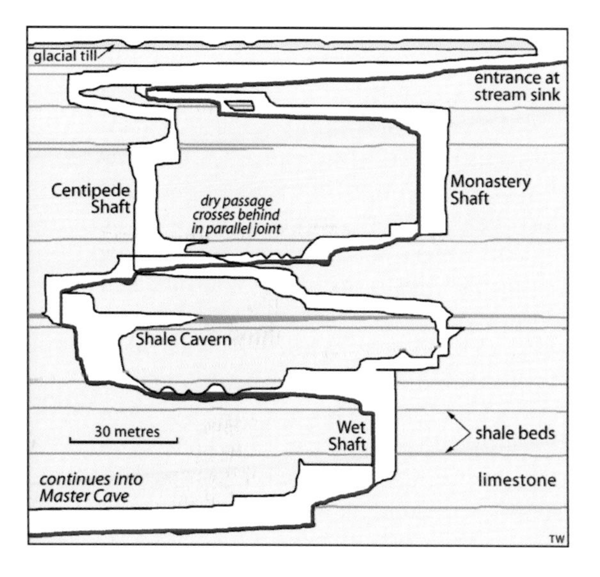

Profile through the staircases of shafts and streamways in the entrance series of Lost John's Hole, under Gragareth; most of these passages and shafts are on a set of parallel joints, and only the active stream route and one of the abandoned routes is shown.

Dales Caver

Dave Brook has all the qualities of a Dales caver – he is slim, wiry and just a little bit crazy. He lives out in Leeds, but for years he has spent nearly every weekend probing the caves of the Yorkshire Dales.

His major interest has been surveying the caves, making the high-quality maps that explorers and scientists rely on. In doing this he has searched every nook and cranny to record the details of the cave morphology, so that he now understands the caves, their geological structure and how they were formed. He is a textile scientist and not a geologist, but he knows the Dales limestone better than most, because he has seen so much of it – from the inside looking out. He knows well that caves are unpredictable, but he knows them so well that he has been among the most successful of Dales cave explorers – discovering new passages that everyone else had just missed. Dave was at his most formidable when teamed up with his brother Alan; and they shared in some of their finest discoveries.

Dave Brook underground.

Beneath the western limestone benches of Kingsdale, Swinsto Hole is a splendid stream cave. It was explored a long time ago, through a long, low and wet crawl, then down seven waterfall shafts and along some fine rift passages with cascades into pools, to 'end' in a tall, boulder-floored chamber. It was one of the Dales' classic caves, frequently visited simply for its sporting delights. Only many years later did one of Dave's friends poke about in that boulder floor and find a very low, very wet passage leading off underneath a waterfall. He could not return to follow up his discovery, so he told the Brook brothers to check it out. Early next year, Alan was first to go under that waterfall, as he had the better light, but Dave was straight behind. And they explored along a bedding plane passage until they made their great breakthrough – they arrived on the banks of Kingsdale's underground river.

This was exploration at its most exciting, and they ran first downstream and then upstream, to take the first-ever footsteps through the clean-washed tunnels of the Kingsdale Master Cave. That summer they returned to the cave almost every weekend; Alan and Dave will be the first to declare that caving is all about teamwork, and they were joined by numerous friends to explore many kilometres of previously unseen cave passages. Roaring streamways, muddy abandoned tunnels, decorated chambers; they were all there, and Dave went to the end of every passage in making his survey. Surveying makes for slow and very long trips, and each trip required an assistant surveyor. Often it was Alan. Eighteen hours into one trip and near the farthest end of the longest passage, Alan suggested that food would be welcome, only to be told by Dave that 'food is merely of psychological value' as he turned back to his survey notes. Sibling friend-

The streamway canyon of the Kingsdale Master Cave, which Dave Brook explored with his brother Alan.

ship was stretched just a little bit far that day. But the cave system under Kingsdale is still regarded as one of the finest in the Dales, and so is Dave's map of it.

A year later, Dave was itching to discover a new cave, so he took the geological route. He set off for the day to walk the shale boundary round the summit mass of Ingleborough, checking out the many small stream sinks at the edge of the limestone. Almost at dusk he found a small open pothole on Black Shiver Moss, above Chapel-le-Dale. An easy climb down brought him to a bedding plane passage with just a few centimetres of airspace over a pool. He had to move a few blocks to be able to crawl in, into a higher passage – all of 30cm (12in) high. Then a bank of mud and shingle required a bit of serious thought – the open passage was only about 19cm (7in) high. But Dave is very thin. He squeezed through, and found a passage that became steadily taller, until he was stopped on the lip of a waterfall 12m (40ft) deep. The next weekend he returned with Alan, cleared some of the shingle to make the entrance crawl slightly easier, and carried in wire ladders to explore further. Black Shiver Pot proved to be a magnificent cave system, where a vertical shaft 80m (260ft) deep is the highlight of a journey that follows a small stream into the depths of the limestone.

Dave does less caving these days, but he is followed by a new generation of cavers who still discover new cave passages in the limestone of the Yorkshire Dales – and who achieve that special thrill of treading where no man has trodden before, right in the heart of a national park visited by thousands.

A small part of Dave Brook's map of the Kingdsdale Cave System; the drawing is the original (with standard cave survey symbols), but colours have been added. Roof heights and water depths are Dave's original figures in feet (about 3ft = 1m). The downstream flooded passage has been added from maps by cave divers.

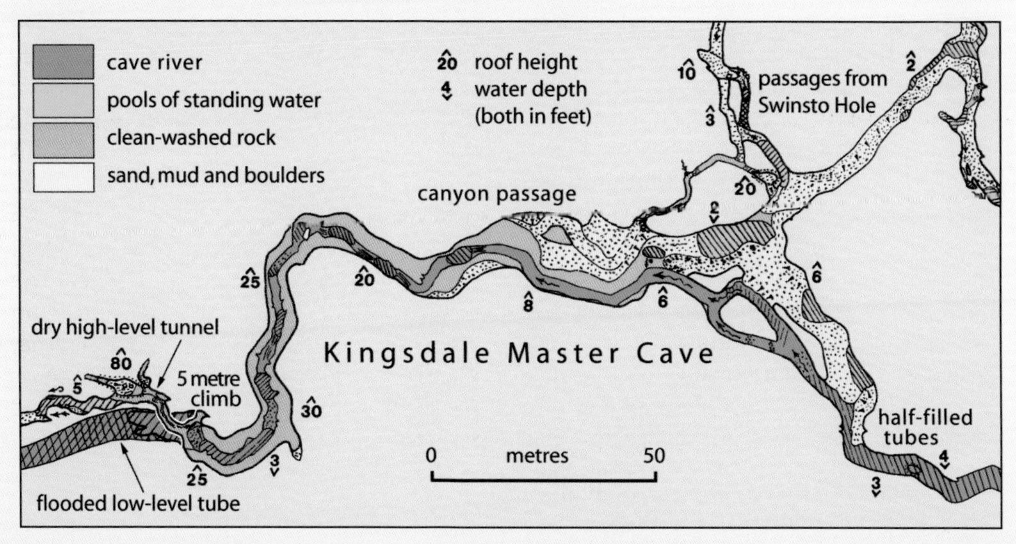

Near the floor of a 60m (200ft) deep shaft in Juniper Gulf, where the whole cave has developed down a major joint in the limestone beneath Ingleborough's eastern benches.

Joints almost vie with the shale beds to carry cave streams through the limestone. Inside the caves, some streams descend whole chains of joint-guided shafts. King Pot and Swinsto Hole, under the benches on opposite sides of Kingsdale, each have eight waterfall shafts in their descents to valley floor level. In contrast, Long Kin West, on the southern slopes of Ingleborough, descends 168m (550ft) in two great shafts that lie almost on top of each other within a single cluster of major joints.

A handful of caves in Wharfedale are noted for their long straight passages, which formed along individual major joints. East of Kettlewell, the headwaters of Dowber Gill Beck sink into an inauspicious little cave, which winds down into the Dowbergill Passage. This continues northwards in a dead straight line for 1400m (4600ft) until it joins the streamway of Dow Cave above its exit into Caseker Gill. For most of its length, this straight passage is over 20m (60ft) tall and less than a metre wide, with the stream on its floor, and fallen blocks and old flowstone creating false floors at various levels. Just one major joint in the limestone has directed the underground stream in a direction at right angles to the surface slope. A little further up Wharfedale, the Birks Fell Caves lie under the opposite flank, and also have long straight passages that drain almost parallel to the main dale instead of towards it. Whereas the Dowbergill Passage is almost level, the two under Birks Fell each descend over 100m (330ft); shale beds create steps in their profiles, but never manage to deflect the cave far away from the controlling joints.

Outline map of part of Upper Wharfedale, where streams sinking into the limestone high on Birks Fell do not resurge from springs directly down the hillside; instead they flow through long, joint-guided, cave passages to separate risings further down the valley.

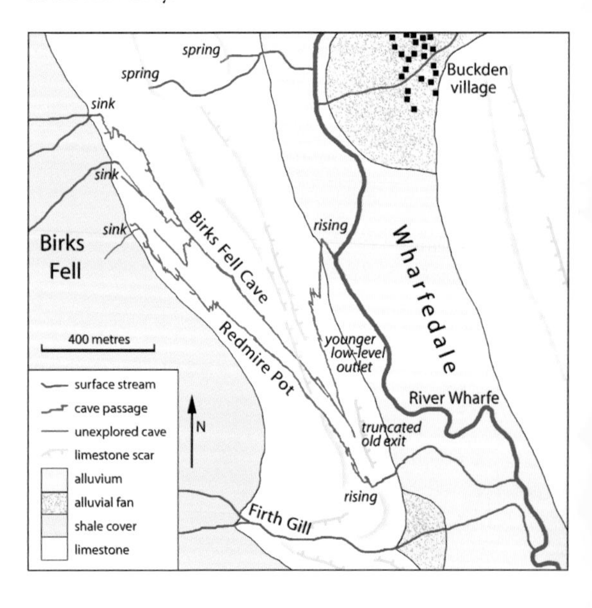

While all the deep and very long Dales caves are in the Great Scar Limestone, there are also some significant caves formed within many of the various limestones of the Yoredale succession. In these thin and nearly level limestones, caves cannot reach to great depths, but some are long. The Middle Limestone thickens greatly in the eastern Dales, so it is here that it contains a few major caves. Just below the sharp bend where Nidderdale turns south, the River Nidd sinks into Goyden Pot and flows underground to the Nidd Heads resurgences at Lofthouse. Part of its route has been explored through some large cave passages that follow the lower beds of the Middle Limestone with sandstones and shales both above and below.

High on Grassington Moor, a large stream sinks beneath the rock buttress of Mossdale Scar, into a cave notorious for its long and rather small streamways, which require seemingly endless crawling for those who venture in. Many of the passages of Mossdale Caverns lie right on the base of the Middle Limestone, so they have triangular cross sections well under a metre high and wide; the floor is smoothed sandstone and the roof rises to an apex along each joint in the limestone. Also in the Middle Limestone, the nearby Langcliffe Pot has long passages that are similar but are mostly of walking size. Its stream can be followed down fault zones where it cuts down through a sandstone level into the Simonstone Limestone, and then through another sandstone into the Hardraw Scar Limestone. The latter is continuous down into the Great Scar Limestone, but the cave cannot be followed further though a major collapse zone. Both Mossdale and Langcliffe drain to the Black Keld rising on the floor of Wharfedale, and it is likely that the Mossdale water also finds faults in order to step past or flow through the non-cavernous sandstones.

Most of the numerous small caves in Wensleydale and Swaledale are formed in the Main Limestone. High above the village of Sedbusk, in Wensleydale, some of the headwaters of

The entrance to Goyden Pot, with the River Nidd pouring into it in a state of moderate flood.

Dales Show Caves

There are just three show caves with full tourist facilities within the Yorkshire Dales, but there are also three more sites that are notable for very easy access. All these offer the opportunity to reach inside the Great Scar Limestone and see just a little of the underground part of the Dales landscape.

Opening into the side of Chapel-le-Dale, White Scar Cave is a significant tourist site. It is also a geological treat, because it is right on the base of the limestone, so that its entrance passages have a limestone roof and a slate floor. Most of the length of the cave is a tall, narrow streamway that carries a large stream draining from Crina Bottom and the adjacent slopes on Ingleborough. Its walls are broken by ledges that trace the beds of less soluble limestone, and some are draped with active flowstone cascades. A mined tunnel then leaves the streamway with a path up into the Battlefield Chamber. This is a very old piece of large cave passage, now decorated with some fine straw stalactites; it is only known because part of its floor collapsed long ago into the younger streamway beneath, creating a huge boulder pile through which local cavers found a very constricted route.

Also under the hill of its name, Ingleborough Cave is far from the road, and is only reached by a long walk up Clapdale (*see* page 139). In the geological past, its main tunnel was the outlet for the stream that still cascades into the Gaping Gill pothole, but this now takes a lower route to emerge from Clapham Beck Head. The old abandoned passage was then almost blocked by calcite gour barriers, deposited by tiny trickles of percolation water, until these were broken through in 1837. Lakes had formed behind these barriers, with their surface levels still recognizable in the profile of the calcite flowstones that can be seen beside the cave path. Bedding planes and joints were exploited by the cave stream as the passages evolved with various branches and loops, and some splendid bits of cave geology can be seen on an underground visit. Long Gallery is the farthest point reached by the show cave, and it has a classic T-shaped profile. The wide upper section was cut along a bedding when the cave was full of water; partial drainage allowed the canyon to be cut on its floor, and all has been dry since the stream found a new lower route. For more than 130 years, deposition and growth rates have been measured on the large Jockey Cap stalagmite; its growth rate appears to be about one millimetre every seven years. This correlates well with dated material from cores into the stalagmite, and suggests it started forming about 4000 years ago when climates became wetter and processes of solution may also have been influenced by early forest clearances.

Directly beneath the road from Grassington to Pateley Bridge, Stump Cross Cavern was first entered in 1854 by miners who sank a trial shaft in their search for lead ore. Purely by chance, they broke into a complex series of caves, where high level passages had long been abandoned by any streams and were now richly decorated with calcite deposits. Reaching some of the nearer passages and chambers, the show cave is most notable for its sediments. The passages must have been dry for more than the 100,000 years that some of the stalagmites have been dated back to; and it was during the Pleistocene Ice Ages that open entrances offered shelter to reindeer and wolverines whose bones have been found

A corner of the Battlefield Chamber, before the tunnel and walkways made it accessible as part of the White Scar show cave.

The main chamber in Yordas Cave.

within the cave chambers since sealed in when old entrances were blocked by glacial debris.

Though a popular destination for Victorian visitors to the Dales, the single large chamber in Yordas Cave no longer warrants the costs of commercialization. But it lies not far above the road up Kingsdale, and is worth a visit by anyone passing with a good torch. Its entrance is recognizable by the Victorian stonework and now requires a stoop over a mudbank. But it is only a few paces to where the roof lifts into the stygian darkness of the chamber. Follow the noise up to the right and a waterfall can be found lashing down into a side chamber – underground geology in its natural state.

Alum Pot is a well-known pothole on the fells above Selside in upper Ribblesdale. A stroll up to the pothole is hardly rewarded by the tree-shrouded view down into darkness, but another cave just up the fell is worth a visit. Beyond the stream entrance to Diccan Pot, rock outcrops, fenced off in a corner of the field, are scored by the entrances to the Long Churn Caves. Downstream is just for well-equipped cavers, but the upstream passage as far as the waterfall into Dr Bannister's Hand Basin is a walk-in easy enough for visitors who each have a good torch and don't mind wet feet. A clean-washed canyon passage has walls of pale cream limestone fretted by endless scallops. With a few small waterfalls and only a modest stream in dry weather, this is a classic piece of Yorkshire streamway.

Last but not least of the accessible caves is Gaping Gill, high on the limestone plateau below Ingleborough and reached by a fine walk via Ingleborough Cave and Trow Gill (*see* page 138). Gaping Gill is the epitome of a major Dales pothole that swallows a powerful stream. The Main Shaft is nearly 100m (320ft) deep, so it is only accessible on the weekends of the two summer bank holidays, when local caving clubs rig up a winch and offer descents (and return ascents) to passing visitors. Sliding down that hundred metres in a bosun's chair is an unforgettable experience. The deep shaft, the crashing waterfall and the huge chamber are unimaginable from the surface. A stroll round the half-lit chamber, and perhaps a guided trip along a very dark side passage, place a whole new perspective on the Dales geology; this is an underground landscape every bit as special as the surface landscape.

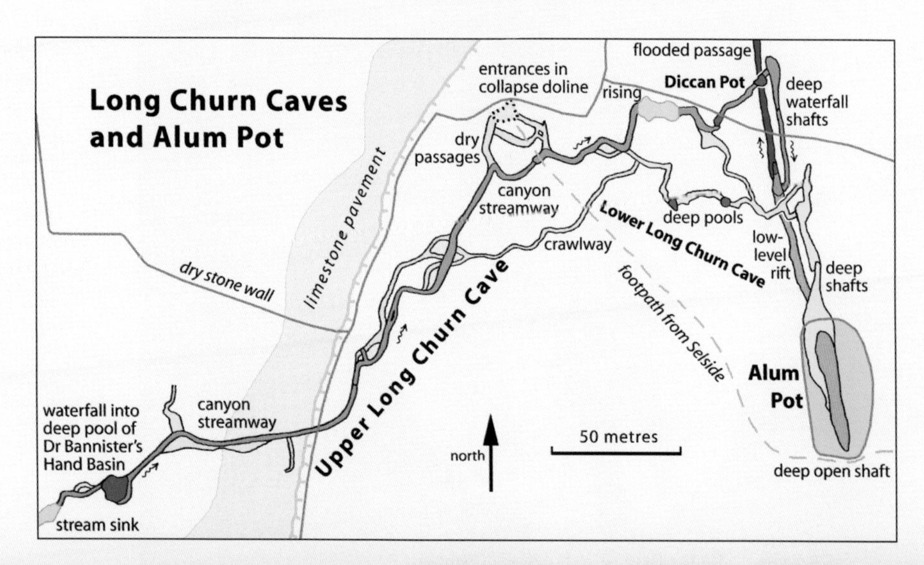

Simplified map of the Long Churn Caves.

Sargill Beck sink into the Main Limestone. But they do not re-appear just lower down the fell; instead, they turn north to follow the gentle dip underneath the surface watershed, to re-emerge from Cliff Force Cave, just below the Butter Tubs, in a valley that is tributary to Swaledale. Cave drainage is controlled by the geology, especially in the layered geology of the Yoredale country.

Flooded caves

Most of the Dales' streamway caves extend downstream into flooded tunnels. Any cave that drains uphill fills with water to its overflow level – which may be the resurgence to daylight, or just a rock lip into another section of open cave streamway. These flooded passages are still parts of the cave systems, but less is known about them because they can only be visited by that small band of cave divers – who rarely have chance to do more than make a map and simply survive in their alien environment of underwater tunnels.

By draining north, down the dip, the vadose caves of the Dales eventually reach the water table. Below this, the limestone drainage is backed up to the level where it can overflow from a resurgence – a pool or stream where underground waters return to the surface. And this is typically where the base of the limestone is exposed in the lower part of a Dale floor somewhere back towards the south and up the dip. Most active phreatic caves, those that are full of water, drop down a joint and then follow bedding planes up-dip to carry drainage towards the main resurgences. These flooded phreatic tubes are mainly elliptical in cross section. Floor, roof and walls are all eroded away, but fastest along any geological weakness, so that the profile extends along any bedding plane or joint that has controlled the

A silhouette in Minaret Passage, deep within the Ease Gill Cave System, clearly shows where the cave passage developed along the intersection of a vertical joint and a horizontal bedding plane.

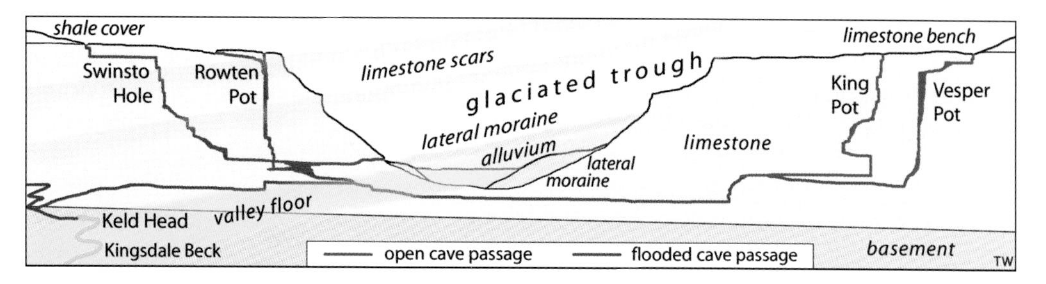

Profile across Kingsdale and just some of the passages in its long cave system. The passage from King Pot, Vesper Pot and many other caves under East Kingsdale passes right beneath the glaciated valley floor, to join the caves from Swinsto Hole, Rowten Pot and many other sinks along West Kingsdale. Downstream from the main confluence, the underwater cave passage comes out of the line of the profile to reach the resurgence of Keld Head further down the valley. The profile is about 2km (1.2 miles) across, and the valley is about 140m (450ft) deep, so the vertical scale is about double the horizontal scale.

passage location. Many beneath the Dales are 5m (16ft) on one diameter, but only a metre or so in the other.

The limestone beneath Kingsdale contains more than 7km (4 miles) of flooded cave passage behind the dark pool that is the Keld Head resurgence. Nearly all of these passages are on just two bedding planes that rise almost imperceptibly to Keld Head and are connected by vertical joints. The main passage drains from the cave system under West Kingsdale, but a major inlet comes from the East Kingsdale caves, forming an underwater passage 3km (2 miles) long through to Keld Head. This passes right beneath Kingsdale, within limestone little more than 20m (65ft) thick, which is all that remains between the glaciated floor of the valley and the slate basement beneath. The rising at the foot of Malham Cove is from another flooded cave that follows a single dipping bedding plane, though what lies beyond the underwater boulder choke 700m (2300ft) in, remains unknown.

Flooded caves beneath the Dales can be much more complex, both in overall route and in cross section, where dissolution has eaten away at a multitude of structures in the underwater limestone. Though a passage formed just below the water table provides the shortest route to the resurgence exit, some flooded caves loop up and down along multiple bedding planes and joints. Deep under the western flank of Gragareth hill, the water from Ireby Fell drains into a passage beneath Gavel Pot by flowing up a vertical flooded shaft 60m (200ft) deep. This is an active phreatic lift, the downstream half of a deep phreatic loop, and one of a series of flooded phreatic loops between stretches of open canyon passage in the long cave system below the slopes of Gragareth.

While active phreatic tubes and lifts can only be seen by the cave divers, many passages formed in this way can be more easily seen in the old abandoned cave systems where the water has drained out into younger caves at lower levels.

Old abandoned caves

The Pleistocene glaciations of the Yorkshire Dales created huge markers within the evolution of the cave systems. During each glaciation the ground was frozen, processes of dissolution virtually stopped, and cave development stood still. But each glaciation also saw the Dales scoured by ice to new depths, and, when the glaciers melted away, the limestone hills could drain to lower levels in the newly deepened dales. So each glaciation was a rejuvenation, not only of the surface landscape but also of the caves below. Many caves that

Walk to Gaping Gill

Most famous of all the Dales potholes, Gaping Gill offers a spectacular underground descent, and also a great walk just to reach its entrance high on the fells of Ingleborough.

From the car park in Clapham village, cross the beck and enter the Clapham Estate past the saw-mill. A delightful walk through the woods of Clapdale is named the Reginald Farrer Trail, after the estate owner of a hundred years ago. He was a great botanist and traveller, and planted the Asian rhododendrons and Himalayan bamboo that still grow in this English woodland. A wide track had already been built up the valley, so that horse-drawn carriages could carry guests to Ingleborough Cave and Trow Gill. It is now an easy path.

Little can be seen of the geology at first, but The Lake is retained by a small dam that stands almost exactly on the South

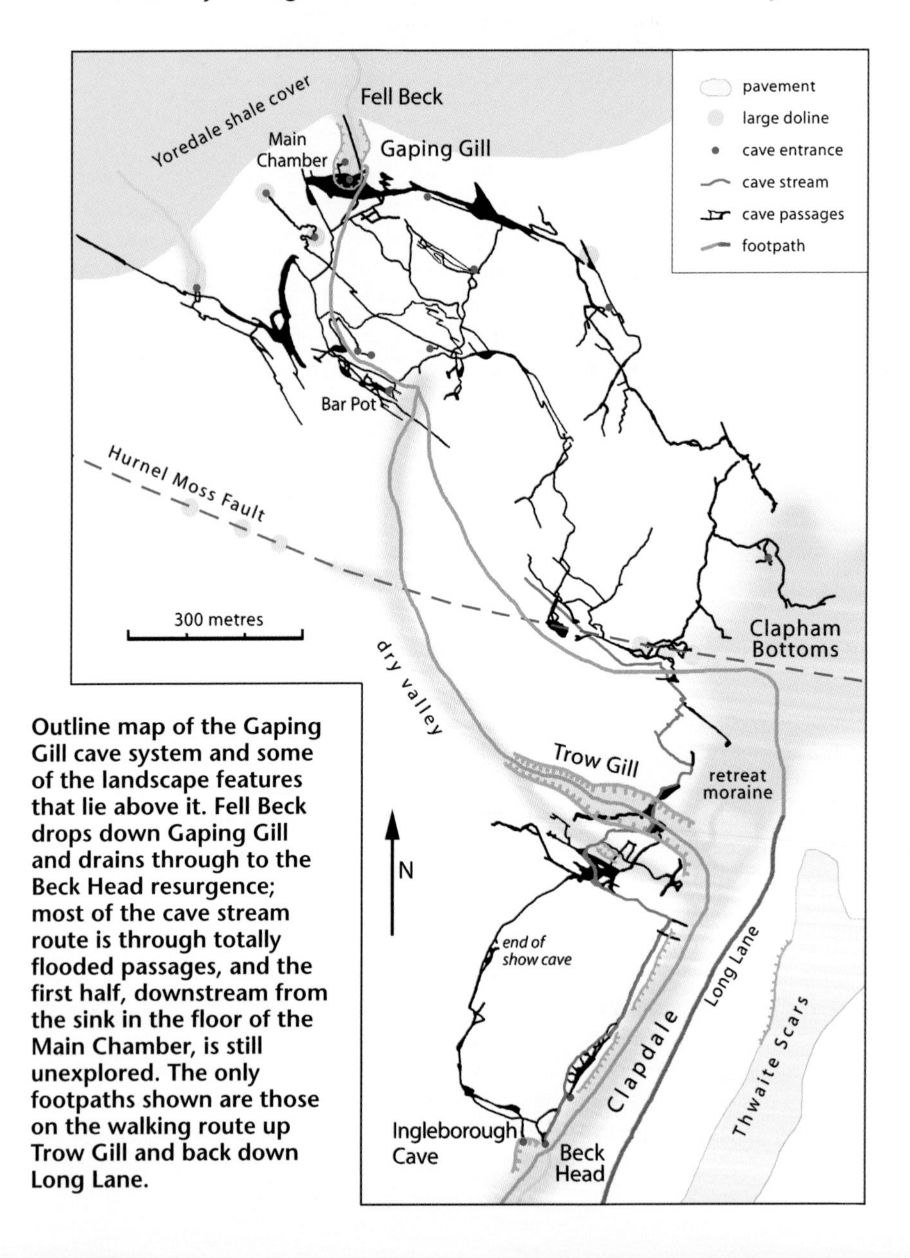

Outline map of the Gaping Gill cave system and some of the landscape features that lie above it. Fell Beck drops down Gaping Gill and drains through to the Beck Head resurgence; most of the cave stream route is through totally flooded passages, and the first half, downstream from the sink in the floor of the Main Chamber, is still unexplored. The only footpaths shown are those on the walking route up Trow Gill and back down Long Lane.

The path to Gaping Gill leads right through Trow Gill, the narrow gorge cut by a meltwater stream, and not formed by collapse of a cave.

Craven Fault. The North Craven Fault is crossed unseen on the rise at the head of the lake, so the next streambed, descending from the flood resurgence from Cat Hole, exposes a tiny inlier of Ordovician slate. Soon the Great Scar Limestone is met at track level, while the powerful spring of Moses Well lies down below closer to the beck and yields the drainage from Newby Moss. Moses Well lies more than 20m (65ft) below Cat Hole, yet both are at the base of the horizontal limestone; the difference is due to the basement relief – Cat Hole sits atop a greywacke hill that was submerged by the Carboniferous sea long after limestone was already forming in a bay where Moses Well now lies.

Through the woods, Clapdale opens out into a wider section as far as Ingleborough Cave. Right beside the track, this is open to visitors, and is well worth a visit (*see* page 134). It is the truncated outlet that once carried the water from Gaping Gill, and largely abandoned cave passages have been followed by cavers right through from sinkhole to outlet. At this lower end, the water has now found a lower route, and emerges from Beck Head, just a few metres up the valley. The next section of Clapdale is lined by limestone scars; it did carry ice during the Pleistocene, but most of its profile is due to erosion by powerful meltwater rivers that could not sink into the frozen ground. Now the stream is in the cave, and the valley is dry.

Just below where Clapdale is almost blocked by a retreat moraine at the foot of Clapham Bottoms, the main path swings left, and up into Trow Gill. Often mistakenly described as a collapsed cavern, Trow Gill is a splendid meltwater gorge – another relic of Pleistocene periglacial conditions when streams flowed on the surface over ground that was frozen solid. The climb up the top end is through a narrow, boulder-floored slot, where wall recesses indicate the positions of deep streambed potholes formed by swirling water; eventually they enlarged and coalesced to form the ravine – a bit like the Strid is today. A meltwater stream cut Trow Gill so quickly because it had the erosive power where it dropped so steeply into Clapdale; and all unrelated to the caves that pass with unbroken roofs nearly 20m (65ft) below the gorge floor.

Above Trow Gill, the path traces a splendid dry valley, with limestone scars showing through the extensive cover of glacial till. Over a stile to the left, Bar Pot is a large rocky pothole just beside the path; it offers the easiest way into the

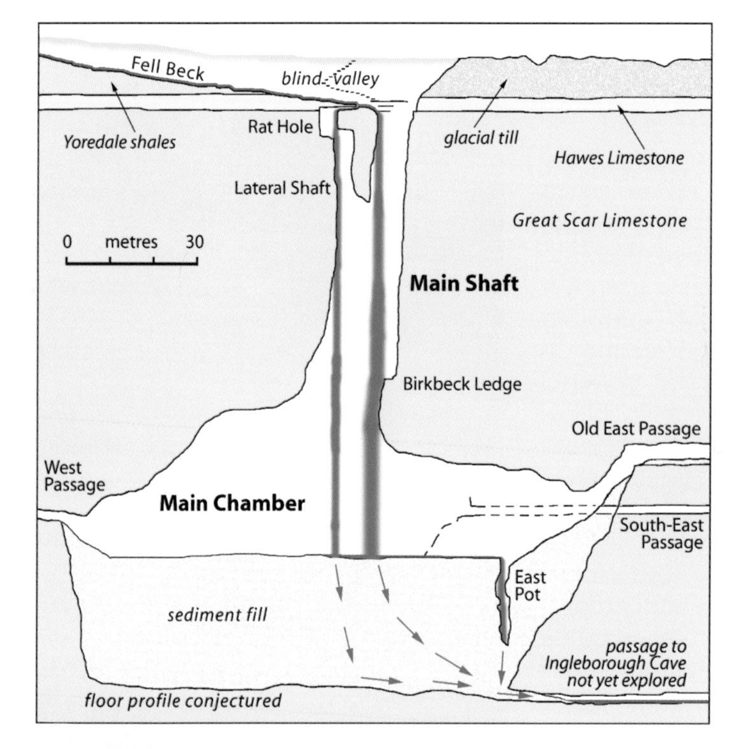

Profile through the Main Shaft and the Main Chamber of Gaping Gill. Beneath the sediment fill, the rock floor profile of the chamber is only estimated from preliminary geophysical surveys. The outlet passage must exist, but its position is only conjectured, as it has not yet been explored by cavers.

rambling cave system of Gaping Gill, but its deep shafts mean that it is can only be entered by experienced cavers. Heavily modified by collapse, the origins of Bar Pot remain unclear in how it relates to the complex of passages and chambers below, all of which are very old and long abandoned by the streams that formed them. Of the various little shakeholes and fissures beside the path beyond Bar Pot, some provide alternative entrances into the Gaping Gill caves, again for cavers only.

Across a flat and boggy piece of drift-covered moor, the approach to Gaping Gill offers no views of the hole short of the very lip. There Fell Beck can be seen draining off the slopes of Ingleborough and disappearing down the great shaft. The beck runs down a blind valley cut 10m

The Main Chamber of Gaping Gill, with Fell Beck crashing down its shaft from daylight, at a time when winch descents would not have been a good idea.

(30ft) into the thick veneer of glacial till, until it meets the limestone and cascades over a rocky staircase of thin beds of Hawes Limestone, the lowest of the Yoredales. Here this is continuous with the underlying Great Scar, in which a series of major joints contain the big pothole itself.

If the winch is in place (*see* page 135), the visitor can follow the joints into the depths, but the rock structure is complex indeed where they merge into at least one fault that forms part of the roof of the great Main Chamber. Both waterfall and winched visitors land on a cobble floor 98m (322ft) below (even though the depth is often cited incorrectly as 110m – 360ft). Three known passages radiate from the chamber, but all are old and abandoned tunnels that are just parts of a complex network of caves. All the water sinks into the debris of the floor, and recent geophysical work has indicated that this debris may be about 30m (100ft) deep, and could be a lot deeper. This indicates that the chamber could actually be over double the size of the open section above the sediment fill. The caves of Gaping Gill are so old, and so many are completely unknown because they are either blocked with sediment or filled with water (or both), that there is still debate as to just how and in what sequence they were all formed. There is still much to learn about the Dales geology.

The return walk to Clapham can follow the same route, but there is a worthy alternative. Over the style near Bar Pot, take the path up the slope ahead, instead of down the dry valley. This leads down into the stream-less, drift-covered bowl of Clapham Bottoms, where shakeholes drain into the caves below; three very large shakeholes beside the path have partly formed by collapse into very old cave chambers, probably along the line of the Hurnel Moss Fault. Up the opposite bank, a gate leads into the head of Long Lane. Fine views of Trow Gill and Clapdale open to the right on a long gentle descent over drift-mantled limestones; then across both Craven Faults, unseen within the final slopes off the Craven Uplands, and back into Clapham village.

The eastern slopes of Ingleborough, with Trow Gill hidden in the trees in a view from the slopes above Long Lane.

The almost perfect abandoned phreatic tube that is the main passage in Sleets Gill Cave.

had previously been flooded were now dry, new stream caves were formed by vadose streams draining to new depths, and another generation of phreatic tunnels developed below a new water table.

Old abandoned cave passages riddle the limestones of the Dales. Many are now reached through younger stream caves that intersect them, a few have been truncated to leave open entrances in the hillsides, but many remain undiscovered behind sections of roof collapse, sediment chokes or even massive stalagmite fills. Most extensive are the old phreatic caves that now lie dry and silent. Beneath the slopes of Ingleborough, the great Battlefield Chamber lies high and dry above the modern streamway and is now reached by the tourist path in White Scar Cave. It once carried the main underground drainage of Chapel-le-Dale, and, because it was part of a great phreatic tunnel, it must have been below the water table, and below the level of the contemporary valley floor. It pre-dates the modern valley, which has since been cut some 50m (160ft) deeper.

Grand chambers are not the only old abandoned phreatic caves. Tall rifts on joint lines were once filled with water, but now stand dry, looming into the darkness above many a cave passage in the Great Scar Limestone. At various sites around the northern Dales, dissolution beneath the water table etched out intersecting sets of multiple joints in the Main Limestone to create bewildering maze caves; these are very old and now lie far above the valley floors as veritable lattices of small passages. Bedding plane passages were also etched out below the water table. Hensler's Passage in the Gaping Gill system is the classic example, 400m (1300ft) long, mostly 5m (16ft) wide and nowhere more than a metre high; it was left dry when the water table dropped and no vadose stream has ever cut a trench in its floor.

Many of the long and nearly level, abandoned phreatic passages probably lay only just below their contemporary water tables, like the Keld Head caves today; but others lay deeper. The main passage of Sleets Gill Cave, in Littondale, is a splendid phreatic tube. But in the down-flow direction it turns up into a steep tunnel that rises 25m (80ft) to where it is truncated in the modern valley side. This was once an inclined phreatic lift, and the main tunnel must have been at least that 25m (80ft) below the water table.

Part of the Sleets Gill phreatic tube does carry a tiny stream today – a vadose stream that has invaded the older passage, and has cut a tiny trench in its floor. A long section of the main passage in White Scar Cave was also once phreatic, but drainage of the limestone has left it as a vadose streamway, where the modern stream has cut a large canyon in the floor of the ancestral phreatic tube. Its cross section is that of a keyhole – very distinctive and a clear example of the evolution of that cave passage.

Stalactites and stalagmites cannot form underwater in the flooded caves. They can decorate the walls of a vadose stream cave, though they cannot grow within the fast-flowing streams. Best places of all for the deposition of cave calcite are the old abandoned passages and chambers. Tens of thousands of years can see the accumulation of endless varieties of dripstone and flowstone, some pure white, and others stained by the reds and yellows of tiny traces of iron oxides.

A keyhole passage in the far reaches of White Scar Cave. The tube that the caver is standing in is the abandoned pre-glacial phreatic tunnel, while the modern stream is out of sight in the active canyon that forms the slot in its floor.

Nearly all the Dales caves have some calcite decorations; only the youngest of the clean-washed stream caves have none. Some passages are smothered in calcite. Easter Grotto, an old high-level passage in the Ease Gill Cave System is perhaps the most richly decorated cave in the Dales, but every caver in the area has his own favourite decorated passage or chamber. Some of the chambers in White Scar Cave have excellent straw stalactites – each one hollow, thin and fragile, like a brittle drinking straw. Hanging forests of straws are the hallmark of many Dales caves because they form more easily in cool climates; they make fine substitutes for the really massive stalagmites that characterize caves only in tropical climes.

BELOW: Stalactites and stalagmites make a lovely scene in Easter Groot, a high-level gallery in the heart of the Ease Gill Cave System.

Hollow straw stalactites, made of pure calcite, hang from the roof of Straw Chamber in White Scar Cave.

Records of the past

For more than a million years, natural erosion has lowered the land surface of the Dales terrain. Each phase of rejuvenation and valley deepening allowed new resurgences to develop at the newly exposed lower levels. And behind each resurgence, the regional limestone water table was virtually horizontal. New stream caves then developed down to the new resurgence level, cutting through the newly drained limestone, while some of the caves that had been previously flooded in the phreatic zone were left high and dry in the vadose zone. Unlike the surface valleys, whose old profiles were destroyed by renewed erosion, most of the old caves survived as high-level passages inside the limestone hills. Quite simply, the caves in the Dales not only retain their own history, but also record the history of the Dales themselves.

Within the Ease Gill Cave System, the very old streamway canyon in Short Drop Cave can be followed down into the abandoned phreatic tubes in Gavel Pot. The altitude of the change from canyon to tube marks the ancient water table – at the level of the contemporary resurgence on the floor of the Ease Gill valley close to the Dent Fault. Here and in various other caves in the Dales, these ancient passage transitions are valuable records of past levels of the dales' floors, little other evidence has survived the erosion that produced the long-term surface lowering.

Unfortunately the cave passages cannot be dated, in order to date those past valley

OPPOSITE: Stalagmites standing on a mud bank make a very lovely underground scene in the Pippikin Hole main passage, part of the Ease Gill Cave System, and are a natural record of the very considerable age of this passage.

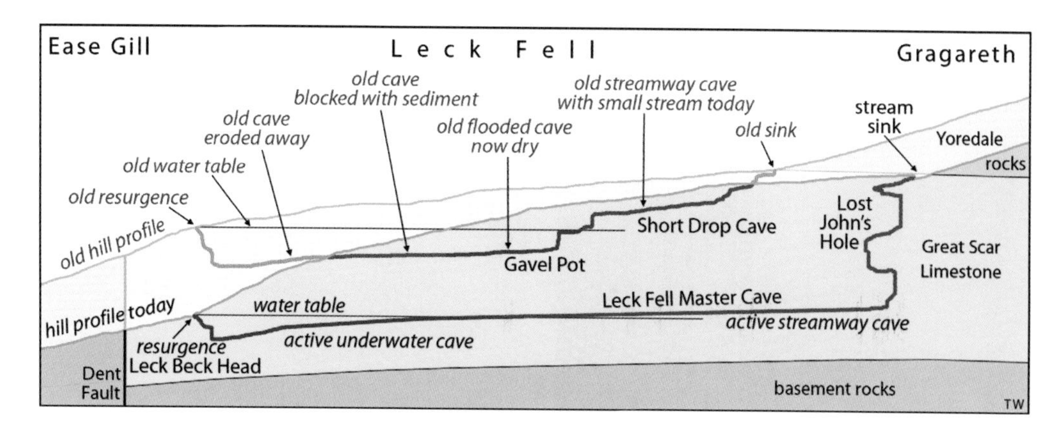

Simplified profile of some of the caves under Leck Fell. Passages in the old cave system of Gavel Pot identify the old resurgence level, and this allows an old valley profile to be interpreted – probably dating to nearly half a million years ago, just prior to the Anglian glaciation. The profile is about 2km (1.2 miles) long, and the vertical scale is four times the horizontal scale.

levels. But the ages of cave sediments can be determined, and they must post-date the cave passages in which they survive. Analyses of the trace quantities of radioactive isotopes of uranium in stalagmites are the key factors in dating the sediments in the Yorkshire Dales caves, though newly developed methods to date either the sand grains or the clay minerals in cave sediments should yield more results in the future. Already, dated stalagmites from the Ease Gill, Kingsdale and White Scar caves have shown that valley floors have been lowered at mean rates of around one metre in every 5000 years over the last third of a million years. The periodic glaciations of the Dales within that time mean that this erosion rate has not been constant through the Pleistocene. However, more detailed data are needed before the relative roles of erosion by glaciers and interglacial rivers can be quantified.

Caves also provide a record of past climates. Stalagmites ceased growth when the ground

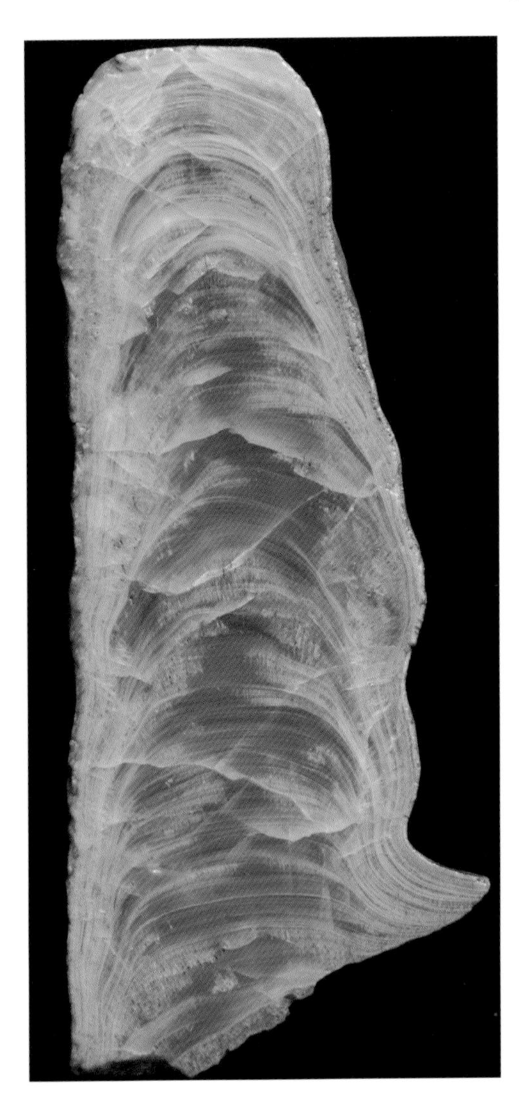

A small stalagmite from a hidden ledge above the streamway in White Scar Cave has been sampled and sliced in two. Its internal banding is a record of changes in past environments both inside the cave and on the fells above.

was frozen during each glacial phase. And when they were growing, ratios of the different isotopes of oxygen within the calcite were determined by the extent of vegetation cover, and hence the climate, at the time of deposition. Stalagmites in the Ease Gill and Stump Cross caves have thereby provided detailed records of climate changes over the last 100,000 years. When drip-water does enter a cave, the stalagmites grow by deposition of extra layers, so that their internal structure can be banded like tree rings. Each ring records the change from summer to winter conditions in the soil and plant cover above the cave, and microanalysis of dated material from Stump Cross Caves has yielded really detailed records of climates over the past thousand years.

Great caves of the Dales

Where active streamway caves converge on to a master cave at depth, and intercept networks of abandoned conduits on their way down, very long cave systems are produced. Ease Gill, Kingsdale and Gaping Gill are the big three in the Dales.

The Ease Gill Cave System is currently the longest known in Britain, with over 70km (44 miles) of mapped passages. Numerous stream sinks on Leck Fell and Casterton Fell all drain through to Leck Beck Head, via two long master caves, one under each fell, though the flooded passages at the downstream end are not yet fully explored. A host of high-level passages include many that are very well decorated, and one of them passes beneath the dry valley of Ease Gill to link the caves under the two fells.

Next door to the Ease Gill valley, Kingsdale is underlain by another great cave system, with deep streamways descending shafts to master caves under each flank of the glaciated valley; these are linked by a flooded connection under the valley floor, and they all drain out to Keld Head. Under West Kingsdale, Swinsto Hole is regarded as the typical Yorkshire cave; it was also the first to be explored right down to its master cave and then out through the flooded passages to its Keld Head resurgence. Between Kingsdale and Ease Gill, there are many more deep potholes, long streamways and large abandoned passages. They all drain east or north to the same two resurgences, and the abandoned high-levels appear to include

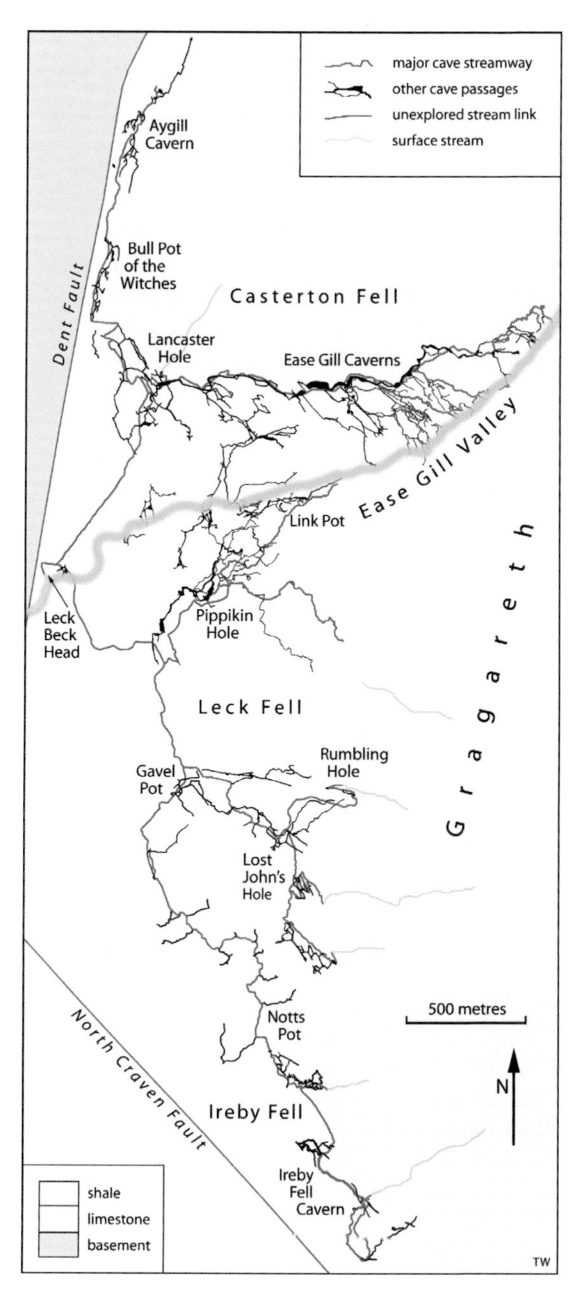

ABOVE: **Outline map of the caves under the western flank of Gragareth. For purposes of clarity, the major cave streamways are drawn wider than reality, and some passages are omitted; 'other caves' include old dry passages, and many that contain small streams. Some of the streamways, and the unexplored stream links, are totally underwater; more passages await exploration.**

passages that once carried all the Kingsdale water through to an ancient resurgence lower down the Ease Gill valley. Some of the old passages are choked with sediment and breakdown, and some of the flooded passages are difficult to follow. But some day in the future, cavers will find ways into the missing links, and a cave system well over 100km (60 miles) long will be mapped beneath the limestone flanks of Gragareth.

Gaping Gill is best known for its deep entrance shaft and its spacious main chamber under the slopes of Ingleborough (*see* page 140). Reaching out from the chamber is a great network of very old, abandoned phreatic tunnels, which are also linked to more than a dozen passages from additional entrances

BELOW: **Brown floodwater, stained with peat, cascades through the main streamway beneath Lancaster Hole, within the Ease Gill Cave System.**

A caver descends a shaft in the entrance series of Lost John's Hole, while a small stream tumbles down the far wall.

scattered across the fell. But Gaping Gill is one of the least understood of the Dales caves. Its largest stream sinks into the rubble floor of the main chamber, and cannot be followed except at the downstream end of its route out through Ingleborough Cave to the Clapham Beck Head resurgence. And all the tributary streams that have invaded the old passages are soon lost into boulder chokes. The ancient abandoned passages present their own problems, as many of them cannot be followed beyond massive chokes of breakdown and sediment. Most of them are roughly level, but that is because they follow the bedding, while a few rise steeply and were clearly ancient phreatic lifts. They were probably formed well below the past water tables, but it is not clear how they were formed or where they drained to many hundreds of thousands of years ago. While the Dales caves are splendid examples of karst development and geological evolution, it is perhaps apposite that the best-known cave in the Dales remains something of a mystery.

BELOW: The large old tunnel at the lower end of Ireby Fell Cavern.

CHAPTER 9

Soils and Plants

Soil is ephemeral; it is just weathered bits of bedrock waiting to be transported by streams and rivers to distant sedimentary basins. Vegetation is just a temporary cover on top of the soil. So may run the views of the geologist who is looking at the long time scale of erosion and surface lowering. Yet soil and vegetation are parts of the one continuous story of landscape evolution. They provide the final details that emerge from a long history of erosion, and because they are at the surface, they are inordinately conspicuous. The Yorkshire Dales are characterized by landscapes that are rooted in the rocks beneath, but they are fine tuned by the soils and plants that, quite literally, colour the view.

Sediments and soils

Most of the soil that blankets the Dales terrain is sediment that has been transported in by any of the natural agents – water, ice, wind or just gravity. It has been deposited as loose material on top of whatever rocks happen to be there, and is therefore largely unrelated to its immediate bedrock.

Alluvium is river-borne sediment that has been spread over valley floors. Clay-rich alluvial soils are deposited where gradients are low – notably along the almost flat floors of the larger glaciated dales – and include various lake fills. This provides the best agricultural land in the Dales – especially the rich meadows that line the river banks in the larger dales. Valley floors with slightly steeper gradients are mantled with alluvium that is mainly sand and gravel, and constitute soils a little poorer for farming. Steeper valleys descending into the glaciated troughs only have the active alluvium that is being moved on by their own streams, but this is so temporary that it barely warrants description as soil. Alluvial terraces

The road across Newby Head, from Ribblesdale to Wensleydale, winds between irregular hills and mounds of glacial till, with a stream bank exposure showing no sign of bedrock.

Autumn berries on a mountain ash make a colourful umbrella over greywacke erratics on limestone scars above Ribblesdale.

are left above floodplain level when a river or stream cuts down to a new lower level; they are not a big feature in the Dales, but terrace remnants can be recognized along parts of upper Swaledale and lower Wharfedale.

Till is the glacial debris that remains from the Ice Ages. Its unsorted and variable soil mantles perhaps too much of the bedrock in the Dales; too much because most of it creates a rather uninspiring and featureless terrain. And because ice once covered the entire Dales, it can be found almost anywhere, though its distribution is largely related to the patterns of ice flow during the last glaciation (*see* page 76). Most of the till in the Dales has a matrix of heavy clay derived from the extensive outcrops of shales that lie between the grits and limestones of the Yoredale sequences and also within the Millstone Grit Series. This creates poorly drained ground, and supports so much of the hill peat that characterizes large areas of the plateaus between the dales, whether their bedrock is grit or limestone.

Though gravity drives river and ice transport, it also creates its own means of natural transport on slopes that are steep enough. Scree, also known as talus, is the most conspicuous result – fragments of rock that break away from cliffs and accumulate in aprons below the exposed crags. Almost every limestone scar has its own basal scree slope, and some of those on the Moughton flank of Ingleborough are textbook ramps of white rock debris. The bulk of these scree aprons accumulated in colder climates of the past, when frost action was breaking more rock from the cliffs above. An active limestone scree offers little footing for plants to establish, but old screes that are now almost inactive are slowly colonized by a plant cover. Cliffs of grit, especially the widespread Grassington Grit, also create scree, but this includes more sand and finer debris, so that plants colonize it faster and it is never as visually striking as the coarse, inhospitable, white scree of the limestone.

ABOVE: Aprons of scree debris lie at the foot of limestone cliffs on Moughton Scar that were plucked by glaciers moving south across the benches (towards the left), while the Yoredale slopes of Simon Fell rise beyond.

BELOW: In a side valley to Wharfedale, terracettes in a bank of grass-covered till have been formed by soil creep, and some are now clearly used as sheep tracks.

Scree is produced on steep slopes, and so are landslides. They are both fallen rock, they just differ in scale. And between them lies a whole spectrum of slope materials, created where large or small sections of ground descend the slopes slowly or rapidly. The whole suite of materials is referred to as colluvium. Maybe little known because, except for the scree and landslides, they are so ordinary, these are hugely important in the formation of soil – and they provide the mantle on almost every slope in the Dales. Soil creep, soil slumps and soil wash are descriptive terms for processes that all contribute to colluvium and soils on slopes. Terracettes – those parallel ledges each perhaps 20cm (8in) high and wide around steep grassy slopes – are initiated by soil creep where the top layer bound by plant roots is gently rucked-up as it slides slowly downwards. Sheep then find that they make very convenient trackways around steep hillsides, and trample them into even more conspicuous ledges; so much so that they are often known as sheep tracks. There is still debate over the role of sheep in forming these soil ledges, but terracettes on slopes throughout the Dales all have a history of sheep activity. Solifluction is the next stage up from soil creep, where saturated soil slumps down slopes rather faster – typically centimetres per year. The ramparts of Arkengarthdale, below Fremington Edge, are patterned by sweeping lobes that are the toes of broad solifluction flows developed in the thick apron of colluvium. These are just the most obvious of many zones of solifluction on the shale slopes between the grit steps in the Yoredale country of the central Dales.

Wind transport is fairly minimal in today's Dales, because it is only effective at moving dry and fine-grained sediment. But windblown silt, known as loess, was a major feature of the barren landscapes left as the Ice Age glaciers retreated. Now it is largely lost within the soil profiles, except on the limestone benches – where it was the only soil-forming sediment outside the areas of till cover. It still forms much of the thin loamy soils that support the grass cover across large areas of limestone, and is notably thicker on some of the slopes around Malham, but no-one knows how extensive it once was. Huge amounts of mud in some of the Dales caves could be loess that once covered much of the pavements – until it was washed down the grykes after man cleared the woodland that once held it in place.

Distinct from the transported sediments, a separate group of soils is produced by weathering of the rocks that are exposed at the surface. So these are directly related to their

A small stream cuts into thick peat hags on the high moors near Tan Hill, exposing broken blocks of hard sandstone. If it wasn't for the nearby coal, this peat could well have been worked as a valuable source of fuel.

immediate bedrock. Grits, sandstones and greywackes all weather to sandy soils, but on the high plateaux of the northern Dales, these are usually masked by peat. Shales and slates weather down to heavy, clay-rich soils, which then provide fertile ground where the shale-rich Yoredale rocks are exposed at lower altitudes, notably in Wensleydale. Limestone produces no soil at all, as it is entirely dissolved in rainwater during weathering, so bare rock scars and pavements identify its outcrops across the southern Dales. However, much of the limestone outcrop is blanketed by transported soils, which support the calcareous grasslands that are the very best for upland sheep grazing.

ABOVE: At the northern corner of Ingleborough, Sleights Pasture is the grassland pocked by shakeholes that lies on the remnants of glacial till between areas of bare limestone.

BELOW: Moors of Millstone Grit around Birk Dale are blanketed by peat and coarse grass.

Wood anemones flower among the coppiced hazel of Aysgarth Wood, growing on a thin glacial till over bedrock limestone.

Peat is its own type of non-transported soil, as it forms by plant accumulation on top of an existing mineral soil, and its distribution is essentially a feature of drainage. So it is widespread on the high grit plateaus of the northern Dales, which receive the high rainfalls. On the well-drained, soil-free limestone plateaus, peat can only establish on any cover of glacial till, though in some places the plant mass can be seen to spread, very slowly, outwards over bare limestone to create a rather dry variety of organic soil.

Grass, flowers, rock and heather

Open grasslands extend across the limestone plateaux to characterize the landscapes of the southern Dales. Blue moor grass dominates the rough land at the highest levels, while lower down there are more lime-loving fescues, thyme and sedges. Flowers are thin on the ground, but the lovely blue harebells can sometimes be found. All new growth falls prey to the constant nibbling of sheep, who are best fed grazing on the nutritious fescue grasses, and also take out the shoots of any encroaching trees and shrubs. The natural cover on sheep-free limestone is an ash woodland rather like that in Grass Wood in Wharfedale. Juniper bushes are spiky enough to keep away sheep, so they thrive on some of the open limestone plateaus, but perhaps the finest are the isolated rowan trees, also known as mountain ashes, which produce glorious crops of bright red berries in September.

Across the southern Dales, the limestone pavements create a niche environment that is a

A hart's-tongue fern safely housed in a deep gryke within a limestone pavement above Chapel-le-Dale.

Grasses, mosses and shrubs are creating their own organic soil that is spreading across the limestone pavement free of sheep grazing in its Scar Close enclosure.

botanical treasure. Their deep grykes provide shelter to a range of ferns and alpine plants, and the broad green leaves of the hart's-tongue fern are especially characteristic. Without any sheep grazing, a pavement can be slowly colonized with lichens, mosses, grasses and numerous small flowers, which are then followed by heather and juniper as the plants accumulate their own organic soil over the intractable rock. A stretch of pavement at Scar Close, on Ingleborough, has been fenced off from sheep for over a hundred years, and is steadily gaining a mature plant cover. Eventually this will progress to woodland, as seen at Colt Park, on the opposite flank of Simon Fell, where ash, rowan and bird cherry create a truly inhospitable terrain of tangled trees over deeply fissured limestone.

Buttercup meadows in Swaledale are one of the seasonal highlights of the Dales, at their best in June, when the whole dale floor becomes a sea of yellow. Buttercups may dominate, but many of these rich meadows house a host of flowers. White pignut, purple wood cranesbill, cream sweet cicely and red clover are usually numerous. Less common are the white meadow saxifrage and various orchids; the latter are now very rare compared to a century ago, when they were sold in bunches at Dales markets. With this suite of flowers and herbs all set in a mixture of grasses – sedges, fescues, bents and ryegrass – these meadows are nationally important for their biodiversity, and those in Arkengarthdale and upper Wharfedale are almost a match for their cousins in Swaledale.

Beside the flower meadows on the dale floor sediments, and the grasslands on the high limestone, the third big type of plant environ-

Buttercup meadows provide a swathe of yellow down the floor of Swaledale in early summer.

Purple heather clads Grinton Moor above Swaledale, except where the old flue from the Grinton smelt mill cuts its line up a hillside.

ment within the Dales is the wet moorland. The juncus reed grass is the hallmark of wet ground, but purple moor grass, cotton-grass and nardus mat-grass are also abundant; any of these can form the tussocks that are so dreadful to walk across. This terrain is typical of many of the high moors on grit on each side of Swaledale, as well as areas of limestone that are mantled by thick glacial till within the southern Dales. Sombre tones are the natural hue of the grit moors – except for those with a rich cover of heather on the slightly more sandy soils. These are beautiful when the heather produces its tiny purple flowers, and the Moors of Grinton and Carperby, on the high ground between Wensleydale and Swaledale, display the best of the colour in August and September. Heather is kept down by excessive sheep grazing, but many areas are nurtured into maturity and continual re-growth so that they can house large numbers of red grouse; sadly this is only economic when managed by gamekeepers for the benefit of sport shooting.

The only ground in the Dales wetter than the grit moors are the mires and bogs. Spongy sphagnum moss dominates to produce the incredibly soft ground, but widespread cotton grass adds flashes of white across seas of dark. Blanket bog that is a mixture of heather and cotton grass has more variety and may be rich with bilberry or crowberry, and also cloud-berry on the higher fells around upper Swaledale.

Grit and shale under the lowlands do not produce such distinctive plant regimes – though this is partly because so much has been

A host of flowers colour a conserved herb meadow in upper Wharfedale.

modified by farming. Strid Woods, in lower Wharfedale, is a delightful exception with its splendid stand of mature oak. And there is one other plant that is notable for being unwelcome. Well drained slopes on colluvium ramps along the dale flanks are easily invaded by bracken, which is useless to any animals and has spread all too fast over the last hundred years or so. As a natural invader, bracken has changed the nature of parts of the northern Dales, especially around the areas disturbed by the old lead mining, and it sits alongside the various changes to the Dales' landscapes made directly by man.

Away from the natural

More than any other aspect of the Dales landscape, it is the plant cover that has been most changed by man. Natural limestone grasslands are becoming rarer as they are replaced by intensively managed pasture that is more productive to agriculture. Meadows are artificially re-seeded to become richer in rye grass and poorer in species diversity. Most of the original blanket bog has been degraded due to inappropriate burning, moorland drainage and atmospheric pollution, all of which has reduced the cover of sphagnum moss. These are only some of the more subtle changes that have been creeping in through the last two millennia.

But farming is an integral part of the Dales, and it creates much that is good in the landscape. Open grasslands have been a feature on the limestone since before sheep numbers increased so greatly on the huge monastic farms of the Middle Ages, but they are not the truly natural ground cover. Sheep are now being kept off some areas, which are grazed only by a few cattle, so that they can return to their natural cover of scrub and bushes, and this 'restoration' is funded by government grants. But this is a purist approach to nature, and man and sheep are actually components of nature in its broad sense. Nibbling sheep should perhaps be regarded as an acceptable agent of landscape development if they maintain the grand sweeps of clean grassland that are such a splendid feature of the southern Dales.

In Swaledale, the browns of late winter are old heather and bracken on the fells above Reeth, in contrast with the green grass in the lower meadows on soil improved by liming.

Part III: Imprint of Man

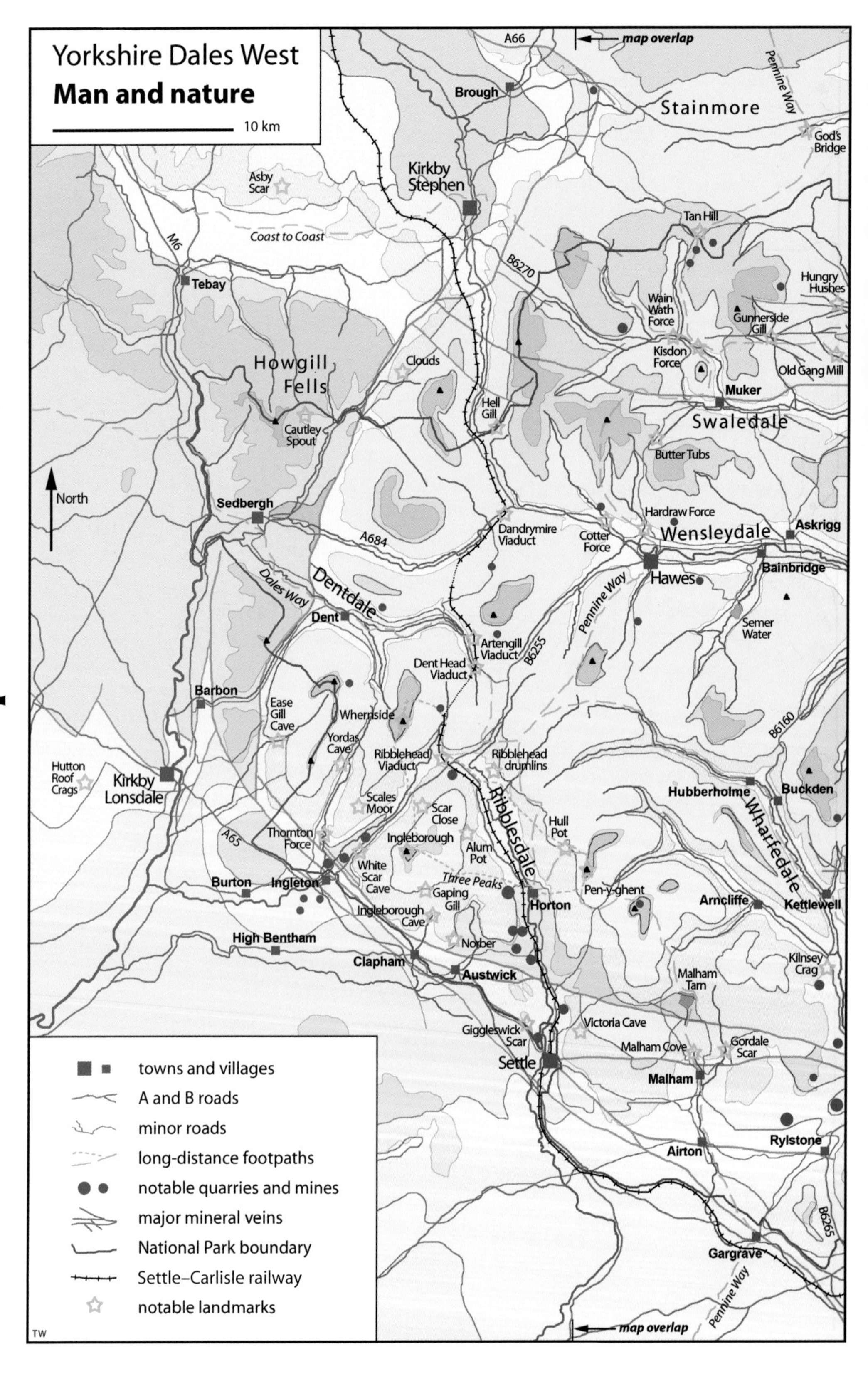

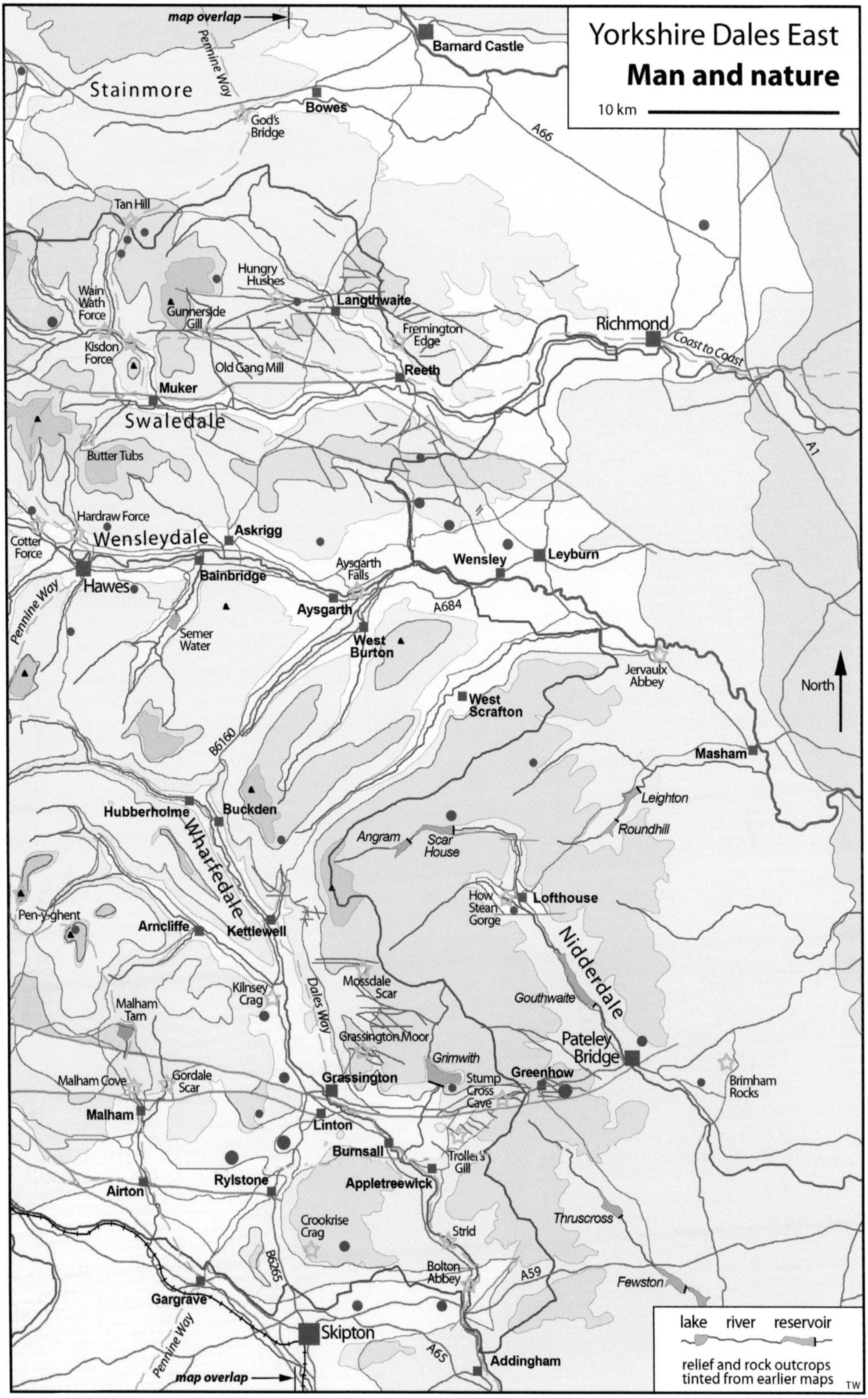

Part III: Imprint of Man

Yorkshire Dales East
Man and nature

10 km

map overlap
Pennine Way
Barnard Castle
Stainmore
Bowes
God's Bridge
A66
Tan Hill
Hungry Hushes
Wain Wath Force
Gunnerside Gill
Langthwaite
Fremington Edge
Richmond
Coast to Coast
Kisdon Force
Old Gang Mill
Reeth
Muker
Swaledale
Butter Tubs
A1
Hardraw Force
Cotter Force
Wensleydale
Askrigg
Wensley
Leyburn
Hawes
Bainbridge
Aysgarth Falls
Aysgarth
A684
Semer Water
West Burton
Jervaulx Abbey
North
West Scrafton
B6160
Masham
Leighton
Roundhill
Hubberholme
Buckden
Angram
Scar House
Wharfedale
How Stean Gorge
Lofthouse
Pen-y-ghent
Arncliffe
Kettlewell
Nidderdale
Mossdale Scar
Kilnsey Crag
Dales Way
Gouthwaite
Malham Tarn
Grassington Moor
Grimwith
Pateley Bridge
Malham Cove
Gordale Scar
Grassington
Stump Cross Cave
Greenhow
Brimham Rocks
Malham
Linton
Burnsall
Troller's Gill
Airton
Rylstone
Appletreewick
Crookrise Crag
Strid
Thruscross
B6265
Bolton Abbey
A59
Gargrave
Fewston
Skipton
A65
Addingham
map overlap

lake river reservoir

relief and rock outcrops tinted from earlier maps

TW

CHAPTER 10

Mining Wealth

An area of limestone and hard rock, such as the Yorkshire Dales, is typically rich in valuable resources of mineral and stone. Today's stone aggregate industry is hardly beautiful, but it is a part of the Dales. Yesterday's mineral industry had a far greater impact. It shaped development in the area, determined the sites of villages and towns, even shaped the hillsides. So, even now, when its activity has declined to zero, it remains a major element within the texture of the Dales landscapes.

Minerals and veins

Until about 1900, galena was almost the only mineral worked in the Yorkshire Dales. Galena is lead sulphide, and contains about 86 per cent lead by weight. It is a mineral, but only becomes an ore when it occurs in large enough masses to be economic to mine. Crystals of galena scattered through granite are just not worth the cost of extracting them. But a vein – a sheet of galena and other minerals filling an opening on a rock fracture, and containing maybe 10 per cent lead overall – is an ore worth mining. Creation of such an ore body requires special geological conditions, but these did pertain in the Yorkshire Dales, and there are hundreds of veins that have been mined.

Most of the Dales veins formed in faults through the Carboniferous rock. A typical vein is nearly vertical, less than a metre wide and over a kilometre long. Veins take their name from an ancient belief that they were parts of a circulation system in the living planet, which is how the Earth was once perceived; fluid circulation they were, but without the concept of life, and the name has stuck. Some veins are known as rakes, an old miners' term more commonly used in Derbyshire. Within a mineral vein, the silvery grey galena occurs within a mix of white minerals known as gangue; these are mainly calcite (calcium carbonate), witherite (barium carbonate) and baryte (barium sulphate), and were worthless to the old miners. The minerals commonly lie in bands parallel to the vein walls. Bands of nearly pure galena a few centimetres thick were the miners' targets, but they had to dig out a lot of waste gangue, and

A chunk of solid galena from a Pennine mineral vein; below the grey galena with its distinctive cubic cleavage, there are two bands of calcite, and there is a coating of quartz on top.

High above Arkengarthdale the Hungry Hushes are scored by a trellis of old mines and open-cuts that worked the multiple veins in this rich mineral belt.

sometimes, barren wallrock, to follow the precious ore in passages wide enough to get through.

Mineralized veins commonly occur in zones of sub-parallel and branching faults, so the mines that exploited them end up as complex underground networks. The heart of the Dales lead industry was in a swathe of veins more than a kilometre wide stretching more than 20km (12 miles) through Keld and Langthwaite beneath the fells on the north side of Swaledale. Friarfold Vein was the longest of all, with mineral 2m (6ft) wide for kilometres of its length. At the opposite end of the scale, very thin veins, formed not on faults but on lesser joints, were known as scrins. These were not worth the costs of underground working, but many were dug out in shallow surface diggings, and some are still recognizable as small straight gullies across bare limestone pavements including those on Fell End Clouds (on Wild Boar Fell) and on Proctor High Mark above Malham.

All the big veins had relatively rich ore shoots lying between areas that were too lean to work at a profit. A good ore shoot may have a band of almost solid galena only a handspan wide but continuous for perhaps 30m (100ft) vertically and maybe a kilometre or more laterally. These were created where the wallrocks moved past each other on faults that were unevenly curved; the results were patterns of thin lenticular openings between places where the two sides remained in contact – and the vein minerals were later deposited in these variable openings. Most ore shoots follow the stronger beds of wallrock. So the miners around Swaledale found the richest ore by following the Main Limestone and the thick cherts just above it, and also the limestone not far below that therefore became known as the Underset. But most of the strong limestones and grits in the Yoredales housed ore shoots in some part of the mining field. In the Grassington mining field, the ore was in the Bearing Grit (the bottom bed of the Grassington Grit) and locally into the underlying Middle Limestone.

In some places the galena was found in ore bodies that extended horizontally, where beds

Mineral Veins

Man's development has for a long time depended on a supply of metals that he could utilize. From simple Bronze Age tools through to the endless machines of the modern world, these have all required metals, which have been extracted from natural minerals dug from the ground. The rocks of our planet do contain plenty of the vital metaliferous minerals, but most are so widely disseminated that it is not worth the cost to extract them from huge volumes of rock. Simple economics demand that miners search for ore bodies in which the minerals have been naturally concentrated, thereby reducing the costs of separating them from waste rock.

Mineral veins are just one type of ore body. They are not the largest, but they are relatively widespread. Because they are vertical, they have a better chance of being exposed on the surface, so are more easily found. Before about 1900, the majority of minerals were worked from veins (with the notable exception of the beds of coal), and the Yorkshire Dales was a very typical mining field.

Long ago it was recognized that vein minerals were deposited in the rock fractures by fluids that came from below. Hot water was the obvious contender as the fluid, and this type of ore deposit is still described as hydrothermal. But vein formation requires more than hot water; it needs three geological processes to come together in the right environment: it needs a source of the metals, it needs a medium to transport the solutions, and it needs a process to concentrate and deposit the minerals. The geology of the Dales provided all three.

Chief suspect for the source of metals has always been granite. The once-hot magma could contain all sorts of metals, and the veins in many mining fields are clearly associated with granite; Cornwall's tin veins are the obvious example. So it was long suspected that granite lay hidden beneath the Askrigg Block as a source for its minerals. No surprise then when the Raydale borehole hit granite, but some surprise when it proved to be older than the mineral veins. Not the source then.

It is now known that the vein minerals of the Dales originated from sediments that accumulated in deep basins. Muds and clays contain only tiny proportions of metals, but there are huge volumes of them in the Stainmore Trough and the Craven Basin, on either side of the Askrigg Block. When these sediments were buried and turned into rock, water left from their formation was squeezed out; it is known as connate water. This happened at depths measured in kilometres, where heat from the Earth's interior raised temperatures

A narrow mineral vein in weathered Yoredale limestone. Though packed with grey galena, and having little calcite gangue, this vein is less than 20mm (just under an inch) thick, so would not be worth mining unless it lay within a zone of multiple veins that could be extracted as one unit.

Bladed baryte crystals that grew into an open cavity in a mineral vein under Grassington Moor. This piece is about 100mm (4in) long.

to 100–150°C. The connate water was actually a brine, about six times as salty as seawater; this is known because tiny amounts of it survive in bubbles (known as fluid inclusions) within the vein minerals. With this temperature and chemistry, these hydrothermal brines could carry all sorts of minerals in solution as they flowed away from the basin sediments. Lead and other metals were easily transported as soluble chlorides.

Warmed and squeezed, these metal-bearing fluids escaped laterally along permeable beds, and always tended to rise towards zones of less pressure. So from the lows of Stainmore and Craven, they rose into the high of the Askrigg Block. This all happened during the Variscan plate movements, soon after all the Carboniferous sedimentation. The muds were squeezed, but the more brittle limestones and grits were fractured by the same plate movements, and displacements along curved faults created narrow open cavities, especially in these stronger rocks. These open faults were the key conduits for movement of the mineralized fluids.

Minerals were then deposited in these same faults, higher in the succession of rocks, to become the mineral veins of today. This was partly because the ground was cooler nearer to the surface, and minerals came out of solution at the lower temperatures. It was also aided by reaction between the saturated fluids and the host limestone. Many veins are simple cavity infills, but many also involved replacement by solution of the wallrock limestone, notably of course in the flots, which replaced certain more soluble beds. Deposition of the minerals took place many hundreds of metre underground; at that time the rocks that now form the Pennine hills were covered by great thicknesses of Millstone Grit and Coal Measure rocks since eroded away. Indeed the thick shales between the Millstone Grits were a barrier to the upward flow of the hydrothermal fluids; the Swaledale ores are mainly in the Main Limestone and Main Chert beneath most of the shales, while the Wharfedale ores are in the Grassington Grit where thick shales lower in the succession were removed by erosion before it was formed.

Every geological story has its twist. The Greenhow mining field is rather different from its neighbours at Grassington and Swaledale, with more quartz, fluorite and sphalerite (zinc sulphide) within its veins. Most significantly, within its sulphide minerals, the chemical isotopes of sulphur indicate that the veins originated not from basinal sediments but from a magma source. It appear that this mineral field was fed from deep below, perhaps from an unseen granite within the basement with hydrothermal fluids rising up the Craven Faults, which bisect its limestone outcrop.

So there is now a good understanding of the geology behind the mineral veins that fed the Dales mining industry. It is perhaps rather satisfying that the old story of their granite source, though now sorely rebuffed, did have just an element of truth in it.

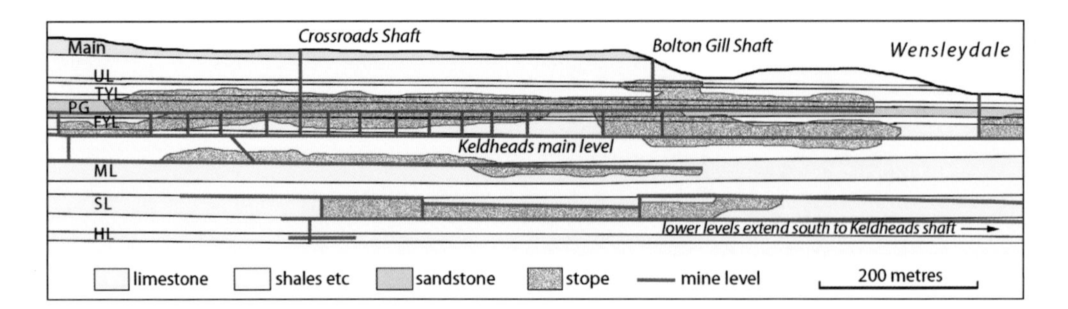

Profile of the Keldheads mine workings along the Chaytor vein. The main stopes match the extent of the richest ore shoots in the wider parts of the vein, which mostly follow in or close to the stronger limestone beds. Names of the Yoredale limestones can be matched with the sequence shown on page 31; PG is the Preston Grit, which is only thick under this part of the northern slopes of Wensleydale.

of limestone had been replaced by new minerals. These were known as flots or flats, and were always off to the side of a mineralized vein. Less extensive than flots were pipes, formed where a mineral vein thickened along its intersection with a particular bed of wallrock, or at the meeting of two veins. Both flots and pipes were welcome bonanzas for the miners, but they were largely found by chance.

Mines and mining

It is difficult to imagine a terrain in England where veins of lead ore could be found exposed on open hillsides. But this was how the first veins were found. All lead ore at the surface was soon exploited, but the veins were followed into the ground, and an industry was born. The Romans worked Yorkshire lead; their date-stamped ingots have been found in Swaledale and on Greenhow, though none of their sites is recognizable through the forests of later workings. Veins around Langthwaite were probably worked by Norse incomers, and then there was an expansion of mineral working directed by the monastic landholders after about 1100. Activity increased in the 1600s, and the heyday of lead mining in the Dales lasted just 100 years between 1780 and 1880. Output then declined rapidly, as the mines came close to exhaustion just before a worldwide fall in the price of lead during the 1880s. None was produced after 1913, but by then the total production from the mines within the Dales was probably close to half a million tonnes of pure lead.

Miners always go after the easiest material first, excavating shallow trenches along the outcrops of any veins that contained visible streaks of shiny galena. These were small open-cut mines, long and narrow where they followed a single vein, but limited in depth where the exposed and weathered walls became dangerously unstable. Without large modern machinery, their excavation was difficult and laborious. So in many of these that lay on steep hillsides, progress was aided by letting streams scour the mineral out; this version of hydraulic mining left the great hushes that are such characteristic landmarks in the northern Dales.

Surface mining soon became impossible, and the miners took recourse to digging out the ore in long and deep underground mines. Access was by shafts and levels. Shafts are vertical, to allow haulage of ore by rope and bucket, mostly 1–3m (3–10ft) in diameter; earlier shafts were dug on the vein, but many later ones were sunk in barren wallrock so that the vein could be totally extracted without affecting the shaft's stability. Bell pits were a variation; a shaft was sunk perhaps only 10m

Hushes of the North

Rocky ravines etched into Daleside slopes – streamless, strewn with broken rock and distinctly lacking in mature vegetation cover, they are suspiciously straight and parallel to each other. They are visible features in the landscapes of Arkengarthdale and Gunnerside Gill; but they are not natural. They are the hushes – left by the miners of yesteryear.

A hush is scoured out of the ground by water erosion – but the water flow is one that has been engineered and directed by the Swaledale miners for their own specific purposes. For each hush, the miners built a small dam of turf and soil on a bench or plateau above any hillside that gained their interest. When water had filled up behind the dam, it was released as a torrent that rushed down the hillside. The purpose of this related to which of the two types of hush that it was – the exploration hush, or the extraction hush.

Exploration hushes were used to scour away the soil and any loose rock, so that clean bedrock was exposed – and any mineral veins could be seen. These were the smaller hushes, because they were redundant once a new vein had been exposed, and most are now difficult to identify with certainty. Among the Hungry Hushes, west of Arkengarthdale, Damrigg Cross Hush is one exploration type that is still recognizable between the larger and later diggings.

Much larger were the extraction hushes, each of which followed a single vein down a hillside. These were effectively an early form of hydraulic open-cut mining, where running water did some of the work involved in rock removal. Once the mineral vein was exposed along its length by washing away the soil cover, hammers, chisels and crowbars were used to loosen both the mineral ore and any adjacent bed rock – especially as the ore was commonly in systems of multiple parallel veins. The dam above could then release another flood pulse to wash the broken ore and rock down the hush, before the whole process was repeated. The water flow down the hush even started to separate the heavy lead minerals from the lighter rock, reducing the effort needed in the mineral separation plant at the foot.

The large old extraction hushes have become permanent features of the terrain, especially along the major belt of veins across the northern tributaries to Swaledale. The largest single group is the Hungry Hushes, west of Langthwaite on the slopes rising from Arkengarthdale to Great Pinseat. Among these, Stodart Hush is perhaps the finest, with its great wall in the Main Limestone capped by the Main Chert. It is difficult to know just how much of this great defile was scoured out by the hush floods, as the traditional hushing was nearly always followed by a phase of more traditional rock excavation of an open-cut that reached deeper into the vein. The hush-top dams have gone, and the hush floor is now buried under debris weathered and fallen from the high walls. The hush is beginning to look like a normal valley; but it has no stream, and it will always show itself up by its alignment on the bedrock geology and not in with the local drainage pattern.

Hushes were a key part of very efficient mining operations, but they were limited to hillsides on the veins, and eventually had to give way to underground mining in order to follow the vein. Most of them date from the years between about 1750 and 1850. Now long silenced, they remain as a tribute to the ingenuity of the Swaledale miners of yesteryear, and have weathered time to reveal yet another aspect of the geology behind the Dales landscape.

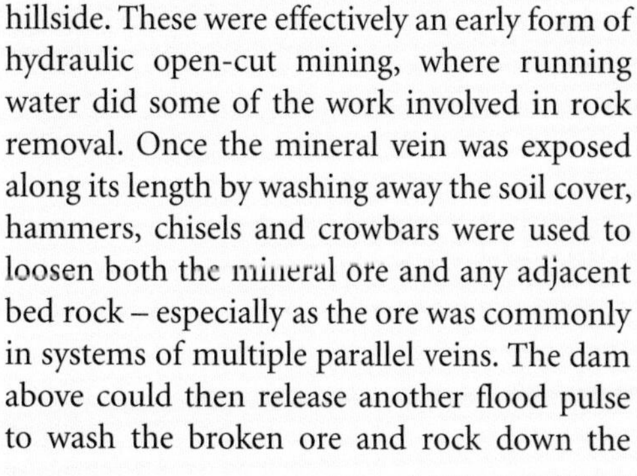

Turf Moor Hush may look like a normal dry valley down the side of Arkengarthdale, but it was all dug out and washed out by miners chasing a rich vein.

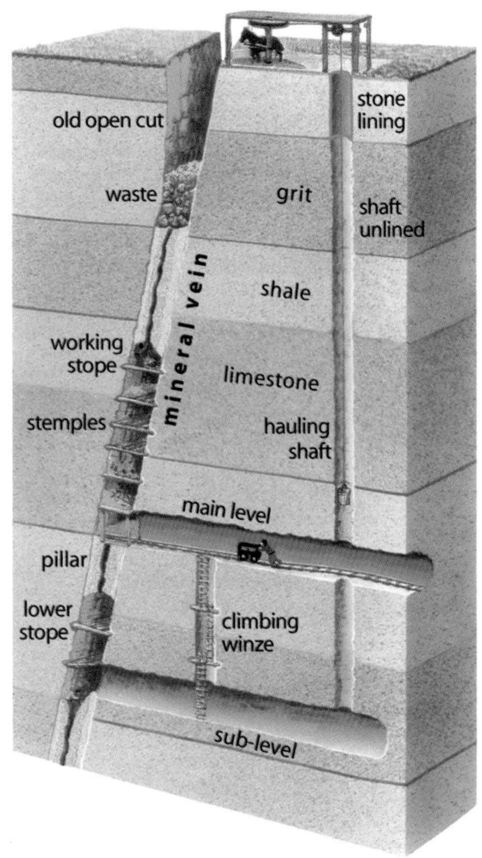

Cutaway diagram of workings in a small Swaledale mine.

(30ft) deep and the ore was then worked outwards, until the roof became unstable beneath its bell-shaped profile, and was then abandoned. Most of the bell pits and the deeper shafts are capped or filled today.

Levels are man-sized tunnels that are horizontal or very slightly uphill to allow free outflow of water. They were driven away from the shafts at various levels, hence their name. But levels were also cut in from valley sides or valley floors, to reach rich ores in veins through beds high within the hillside. These were longer than shafts were deep, so took greater effort to cut, but this was repaid by much easier ore transport in little rail wagons pulled by horses or pushed by men. Some were horse levels purely for ore transport. Others were drainage levels, also known as adits, which were cut in from nearby valleys to prevent the deep mines filling with water and thereby save on expensive pumping up shafts. Many fulfilled both objectives, and one in Lower Punchard Gill, off Arkengarthdale, was a canal level with water ponded enough to allow ore transport in small boats.

Shafts and levels were linked to form networks of galleries, which extended for many kilometres in the larger mines. Branches from the levels reached all parts of the veins,

Emerging from an abandoned mine adit in the valley by the Old Gang Mill.

ABOVE: Water from the old mines of Grinton Moor still drains out of the Cogden Gill Deep Level, which was driven purely to lower the water table on the veins.

BELOW: A vein in the Devis Hole Mine above Grinton has been opened into a tall stope, with a level along its floor that skirts a shaft to a lower level.

with winzes driven upwards and sumps driven downwards, both either vertically or obliquely. Also important were cross-cuts, driven away from the veins. Many were driven for access to veins already known via shafts, but others had the sole or added benefit of searching new ground for unseen veins.

Shafts, levels and all their variants only gave access to the ores. The business areas of the mines were the stopes – the miners' term for sizeable openings created by removal of the workable ore. On veins, these ended up as narrow cavities, vertical or nearly so. In the old Dales mines, some stopes are 30m (100ft) or more high and can reach along for a kilometre or so, though generally broken at intervals by sections of rock left in place, partly to hold the walls apart. These are effectively horizontal pillars, and they also helped provide access. Stopes were generally cut upwards, so that the ore would fall away from the face, while the miners worked from wooden platforms supported on timber 'stemples' that were jammed across the stope. Vertical walls in strong limestone did not need extensive support, so there was little that compared with the forests of coal miners' pit props holding up wide flat roofs in weak Coal Measure rocks.

Most veins are so narrow that some barren wallrock was also broken out to create a stope wide enough for the miners to work in. Many are less than a metre wide. Very welcome were the wider veins, where stopes could match their full width and take only good ore. A good vein could yield 10 per cent galena, while miners worked to a cut-off value of about 3 per cent. If the galena averaged less than this on the whole stope, including any barren wall-rock that had to be taken, then the vein could not be worked at a profit, especially in the deeper mines. Stopes were horizontal in flots, but these were generally not extensive and were patchy in their value; roof support was provided through leaving in place many pillars, preferably composed of rock that would yield less lead.

Excavating hard rock by hand was a seriously strenuous activity for the early miners. Hammering wedges into hand-drilled holes was slow and laborious. Fire-setting was the alternative, where rock at a face was heated by a fire built against it, then chilled by water thrown from buckets. Thermal contraction split the rock, while the miners almost choked in the clouds of smoke and steam. Only since the 1730s did gunpowder ease the task of the Dales miners. And rock drills powered by compressed air were only introduced in the 1860s, too late for most Dales mines then in their declining years.

The restored hillside flue up to its chimney above the old Cupola Smelt Mill, above Grassington.

Mills and smelters

Still inside the stopes, the miners separated good ore from waste rock – not such a hard task with galena that has a metallic shine and is also nearly as heavy as lead. Waste rock, the 'deads', was left, often stacked on redundant platforms, while good ore, the 'bouse', was sent out of the mine. Outside, the 'buckers' (often women and children) used hand hammers to break the ore down to pea-size. Tipped into water-filled tubs, this was jigged or 'hotched' in a crude form of gravity separation – shaken about, the heavy galena sank to the bottom of the pile away from the lighter gangue. The pea-size galena was then ground to sand-size, and further concentrated in 'buddles', where heavy galena was trapped behind baffles over which the gangue was washed. The buddle output was fairly pure galena, ready for the smelter.

In a stone-built hearth, galena was roasted on top of a fire of wood, peat or coal. As the fire matured, the lead sulphide mineral dropped through the fuel, to a lower part much hotter where air was blown in by bellows. At those temperatures, the lead separated from the sulphur; it flowed out of the firebase as molten metal, where it was poured into moulds, ready for sale as the blocks known as 'pigs'.

A distinctive feature of most Dales smelt mills are the flues – stone arch tunnels built up the hillsides, many reaching hundreds of metres long, and originally topped by tall chimneys high on the fells. These flues were truly multi-purpose. They generated a draught through the fire, they allowed lead in the flue gasses to condense on their walls, and they thereby reduced pollution that could be harmful to livestock on the fells. The condensate created their value, and was periodically scraped off the tunnel walls to add to the smelter output. In Swaledale, the Grinton and Surrender Smeltmills have both been preserved, while the Old Gang Mill is more ruined but is impressive in its grand setting. Many of the fell-side flues are still recognizable above the old smeltmills, the best of which has been restored with its chimney on Grassington Moor. Tip-heaps also survive as relics of the bygone mining industry. Those beside the mines are of barren rock and gangue, and are commonly grassed over, while those at the smelters have less plant cover due to their traces of poisonous metals missed by extraction processes that were not perfect.

Mining fields from north to south

Though lead veins could be found scattered beneath many of the Dales hillsides, the good material that was worth extensive mining lay clustered in quite distinct mining fields. There is a coincidental recurrence of the letter G in the locations of the main Dales lead mines; the largest group was centred on Gunnerside

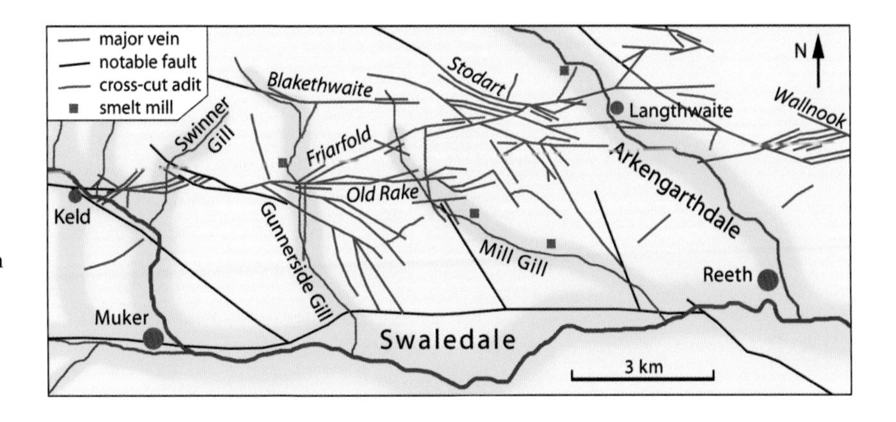

Major veins in the rich mineral belt along the north side of Swaledale.

Mine workings on three major veins score the hillside above ruined mill buildings in Gunnerside Gill.

Gill, and the lead-rich zones farther south lie above Grinton, near Grassington and around Greenhow.

More than 70 per cent of Dales lead came from the rich mineral belt through the hills and valleys on the northern side of Swaledale. These mines were on a great system of parallel and branching veins, and their interconnected workings meant that a miner could walk underground between Gunnerside Gill and Arkengarthdale. The greatest impact on the landscapes was made by the cluster of large hushes and open cuts in the area now known as the Hungry Hushes (*see* page 165), high on the fells above Arkengarthdale; these were most active around the 1750s. Eventually, the miners had to progress into underground workings, and these passed beneath the top end of Hard Level Gill, midway between Gunnerside Gill and Arkengarthdale. Within the period 1589–1913 there were eighteen smelt mills working ore from these mines (and from some others nearby), and they probably produced around 350,000 tonnes of lead from over 550,000 tonnes of galena concentrates. Cross-cuts of no great length gave easy outlets for ore direct from the productive stopes to the Old Gang Smelt Mill, and that one mill

A level in the old Devis Hole Mine, above Grinton, follows the Main Limestone and exposes the black Richmond Chert in its roof.

produced more than a thousand tonnes of lead ingots each year between 1817 and 1878. The ruins of the mill remain as a rather imposing feature in this beautiful and remote valley – today's small memories of yesterday's great industry.

The last working mine in the Swaledale catchment (and the last mine in the Dales with lead as its prime target) was further north near the head of Arkengarthdale. Sloate Hole opened in 1909 on the north side of Faggergill, adjacent to some ground that had been worked in earlier times, but it lasted just five years before its closure.

Grinton Moor and the fells between Swaledale and Wensleydale rank lowest among the four main mining fields. It is however distinguished by the Devis Hole Mine, in Lemon Gill not far above the old Grinton Smelt Mill. Between 1820 and 1875 this worked a series of flots rich in galena within the Main Limestone. It was also notable because its miners got about underground by walking or crawling through some unusual mazes of natural cave passages, which they encountered by chance and then saved themselves from digging some of their easy access levels through the hard limestone.

Second only in lead production to the mineral field north of Swaledale, the huge bleak upland of Grassington Moor lies above a host of rich mines. These produced close to 100,000 tonnes of lead, mainly when they were worked as a single group within the Duke of Devonshire's royalty between about 1790 and 1880. Remains of old shafts, small tips, trackways and smelters lie scattered across the grit moor beyond the tarmac road up to Yarnbury. Smaller mines worked above Kettlewell until 1887, but these have left less impact on today's landscape.

Grass-covered rings of debris surround the capped entrances to bell pit shafts on the veins below Yarnbury, on Grassington Moor.

Lines of shafts, spoil heaps and shallow workings pick out the complex set of workable veins that were found beneath Craven Moor in the Greenhow mineral field.

Straddling the Craven Fault Zone, across Craven Moor and Burhill Ridge, the Greenhow mining field is the most southerly within the Dales. It is different from the others in that it is founded on mineral veins with ore shoots that are largely in two small anticlines of the Great Scar Limestone. Mining started in monastic times and, like elsewhere, the main activity was in the 1800s. Relics of the old mines are especially conspicuous on Craven Moor just east of Stump Cross Caverns, a cave system that was found when miners broke through into its open galleries.

Not just lead ore

While lead has totally dominated mining in the Yorkshire Dales, a few other minerals have proved worth working in the long-gone and more recent past. During the 1600s and 1700s, small amounts of copper ore were extracted from irregular solution cavities in the limestone of Malham Moor. Copper was also won from veins and flats in Carboniferous limestones within the Richmond area, just outside the true Dales. Most of this was won in the 1700s and 1800s, but the Billy Bank Mine, just across the River Swale from Richmond town, worked from 1906 to 1915, and brought the district's total production to nearly 2000 tonnes of pure copper.

Zinc minerals are a very minor component of many of the Dales veins, but are notably enriched close to the Middle Craven Fault. Early in 1788, miners working sparse copper ores beneath Pikedaw Hill, just west of Malham, broke into a series of large, old, natural caves. The floors of three spacious chambers were thick layers of calamine, the carbonate of zinc that is now known as smithsonite. With a newly dug shaft directly into their Calamine Caverns, the miners had an easy time digging out more than 5000 tonnes of zinc ore. The origin of this unique calamine deposit is still open to debate, some of it was stalactitic so appears to be a chemical precipitate, but the dominant powdery sediment may have been eroded and re-deposited by a cave stream. The source of the zinc is also unknown and has led to speculation that larger orebodies, comparable to those in the same limestones in Ireland, could lie hidden at depth. Exploratory drilling in the 1960s was halted while still inconclusive when permission for any mining within the National Park was not forthcoming.

A major gangue mineral in the Pennine veins is baryte, pure barium sulphate that is a notably heavy white mineral. It is valued by modern industry as a component of high density fluids, but was regarded as waste material by the miners of old. Consequently many of the old mine tip heaps contain so much baryte that they are worth excavating and washing to separate the mineral. Dumps at the Old Gang site, off Arkengarthdale, and around Yarnbury, above Grassington, have been reworked sporadically as recently as the 1980s.

Last of the minerals to turn a profit in the Dales has been fluorite, calcium fluoride, still known as fluorspar in many circles. It is an essential component in steel production, and is also valued as a source of fluorine, but veins in the Yorkshire Dales carry less of it than do those in the Pennines to both north and south. Old lead mines on the Burhill Ridge, west of Greenhow, were redeveloped in the 1960s with new stopes supplying mineral to a small separation plant built on the surface. But resources were limited. In the nearby limestone gorge of Trollers Gill, Gill Heads Mine produced fluorite sporadically from 1920 onwards. It finally closed in 1975, when John Russum ran out of workable ore on the vein; he appears to have been the last working miner in the Yorkshire Dales.

After a millennium of mining in the Dales, there are no longer any ore reserves that are proven or probable in miners' parlance. There are only possible reserves of lead ore at depth in the veins of northern Swaledale and in unreached zones below the Bearing Grit of Grassington Moor; add to those the thin possibility of an unseen zinc orebody between Settle and Malham. None has any real promise.

So mining in the Dales is finished for the conceivable future. But the impact remains. Landscapes in some of the mining fields are still dominated by the old surface workings along the veins, notably the great hushes north of Swaledale (*see* page 165). Endless galleries of underground mines lie unseen, abandoned and silent save for the sound of dripping water; they are largely inaccessible beneath capped shafts or beyond collapsed tunnels. Few mine buildings survive, but the ruins of some old smelt mills are now mellowed landmarks on fells that would be very empty without them. The imprint of lead mining also remains in the villages, some of which originated and grew when men could work as miners nearby, creating some of the small communities that have remained within a region and a landscape characterized by scattered farms.

Gill Heads Mine was the last to work a mineral vein in the Yorkshire Dales, when it extracted fluorspar up until 1975.

CHAPTER 11

Stone for Industry

Less dramatic than mines that win sparkling ores of valuable metals, numerous quarries (and a few mines) have long been digging strong stone in the Yorkshire Dales. Supplying blocks of stone for building houses is the craft end of the stone industry; vitally important in yesteryear, this has now diminished under the onslaught of bricks and concrete. But the demand for rock aggregates, to go into roadstone and concrete is always increasing; the stone industry thrives today – it has comfortably outlasted the mining industry within the Dales.

Quarries and national parks do not sit well together, but they are unfortunately both the product of the same hard rocks. The wilderness highlands of Britain are inevitably the resources for industrial stone, and will remain so until there is a national trend towards the exploitation of remote coastal quarries. Environmental controls are much better than they have been in the past, especially with respect to dust emissions, waste dumps and drainage outflow. Impacts are reduced, but quarries can never totally go away; and they do play a significant role in rural economies. In bygone days, stone mines and lime works were major employers, with their large labour forces. Each

continued on p.181

Only the tip heaps at its foot belie the fact that the high section of Stainforth Scar, in Ribblesdale, is not natural. It is a weathered old quarry face, but it lacks the normal benches because it pre-dates regulations on allowable face heights.

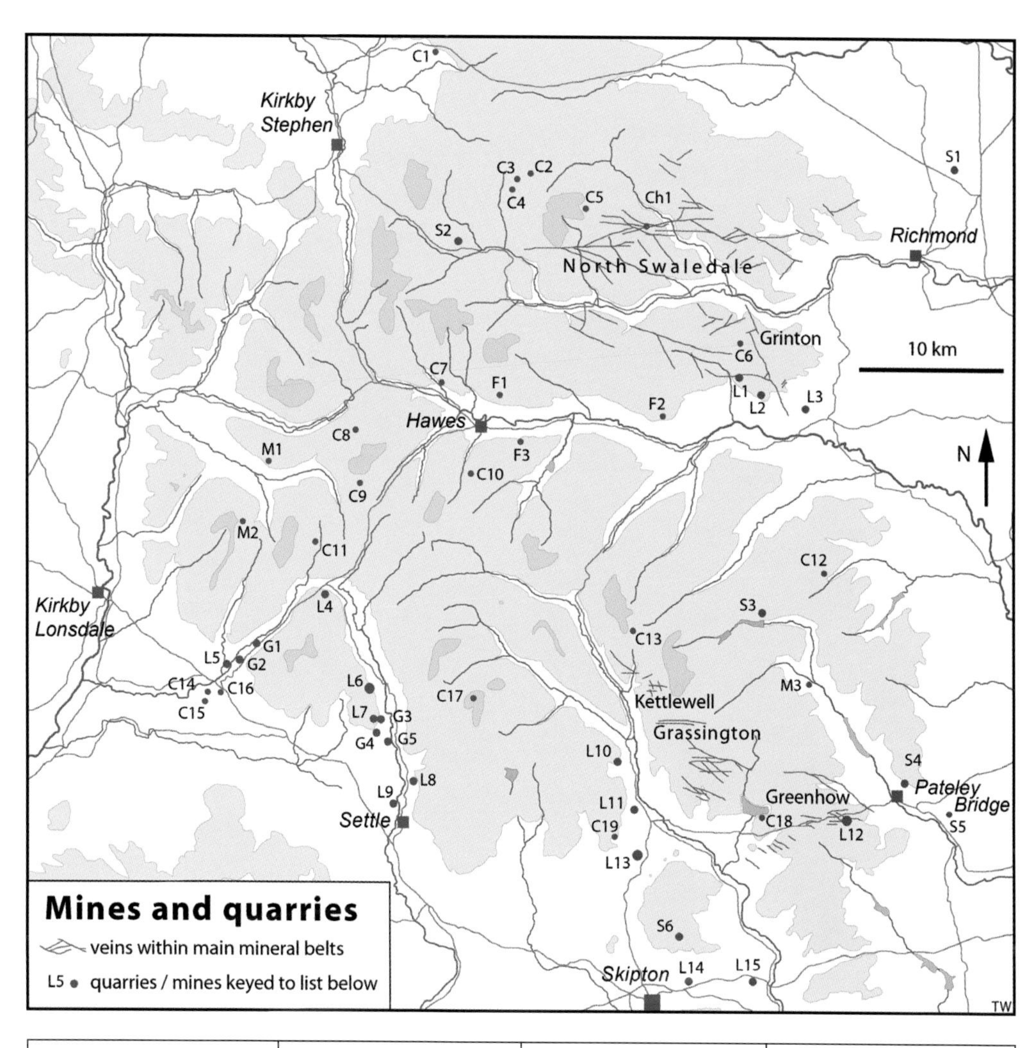

Limestone	**Sandstone**	**Marble**	C4 Mould Gill
L1 Redmire	S1 Gatherley	M1 Scotcher Gill	C5 Gt Punchard
L2 Preston Scar	S2 Hilltop	M2 Binks	C6 Grinton Moor
L3 Leyburn Moor	S3 Carle Fell	M3 Blayshaw	C7 Cotterdale
L4 Ribblehead	S4 Scotgate Ash		C8 Garsdale
L5 Meal Bank	S5 Millstone	**Greywacke**	C9 Cross Pit
L6 Horton	S6 Embsay Moor	G1 Old Granite	C10 Sleddale
L7 Foredale		G2 Ingleton	C11 Coal Gill
L8 Langcliffe	**Flagstone**	G3 Arcow Wood	C12 Colsterdale
L9 Giggleswick	F1 Sedbusk	G4 Dry Rigg	C13 Top Mere
L10 Cool Scar	F2 Carperby	G5 Helwith Bridge	C14 Wilson Wood
L11 Skyrcthornes	F3 Burtersett		C15 Raygill bellpits
L12 Coldstones		**Coal**	C16 New Ingleton
L13 Swinden	**Chert**	C1 Stainmore	C17 Fountains Fell
L14 Haw Bank	Ch1 Hungry	C2 King's Pit	C18 Aket
L15 Hambleton		C3 Tan Hill	C19 Threshfield

Locations of the main mineral belts and some of the major quarries within the Dales.

Dales Quarryman

Robin Gillespie has long had a fascination for stone, and has worked in quarries ever since he left school. Now he runs the big quarries on either side of Ingleborough – at both Ingleton and Horton.

He started at the granite quarry at Shap, working in the laboratories to measure the strength properties of the stone that was being produced to ensure that a good quality aggregate was supplied out to the construction industry. That insatiable need for hard stone to make buildings and roads has given Robin his career. While admitting that production of sand and gravel is of huge economic importance (though only outside the Dales), he feels that gravel pits 'are not real quarries'. Robin has an affinity for real rock. So, within the various companies that have owned the local quarries, he worked his way up the corporate ladder, taking in the quarrying course at Doncaster College on the way. His progress took him to quarries all over northern England, until 1995, when he became established in the Dales.

Robin can feel at home at the Ingleton Quarry, for this is a rock quarry in the best tradition, producing the top-quality greywacke that is so highly valued as a skid-resistant roadstone. It was an exciting time when he started there as deputy manager, because the quarry had an almost new and very efficient crushing plant, and yet the reserves of stone were limited, so the eventual closure was already foreseeable. His first challenge was to plan the workings in order to reach all the reserves and so extract maximum value from both the site and the plant – not easy when the widening quarry left very little space for stockpiles. He then rose to the position of unit manager responsible for both Ingleton and Horton quarries. His new challenge is to manage a smooth transition, whereby Horton would start to produce the greywacke while Ingleton's output declined, in order to maintain the steady supply that is required for road renewal and construction throughout a large slice of northern England.

Currently, Horton is just a huge limestone quarry, and its massive reserves mean that it is destined to eventually be the only working quarry within the

Ingleton Quarry, extracting greywacke from a site squeezed between the Ingleton glens (hidden in the trees) and the road and limestone (on the far side), with slate to both left and right – a management challenge for Robin Gillespie.

The massive Horton Quarry is removing the entire thickness of the limestone along the western flank of Ribblesdale.

National Park. But the greywacke, even more valuable than the limestone, lies directly beneath the quarry floor, and Robin will steer the operation into one that produces both types of rock aggregate from the single site.

Moving into Horton has had an unusual twist, for Robin is practically the youngest man on site. This is a place where quarrying is a local tradition. There are quarrymen at Horton who left school at the age of fifteen and have never worked anywhere else. Only recently, Robin relished the opportunity to present long-service awards to men who had worked for forty-seven years in the quarry. These men give a real perspective to living and working within the Dales.

Robin's affinity for rock extends to a love of rock landscapes. He enjoys being out on the fells, and is fit enough to run the Three Peaks Race each year. So he has been up and down Pen-y-ghent on many occasions, and has to admit that, from that side in particular, the Horton Quarry does have an inescapable impact on the Ribblesdale landscape. But he is pragmatic. The big white limestone scar is already there and will never go away; one day it will even look like the natural scars that are so revered by visitors to the Dales. And the valuable greywacke roadstone due to come from its floor will have significant economic benefits both locally and regionally. Two lots of stone with only one lot of environmental impact appeals to Robin immensely.

The perpetual expansion of modern society continues to demand hard rock to create its infrastructure, and there is always going to be a need for big quarries in hard-rock landscapes. Robin is one of the new generation of quarrymen who really do appreciate their environment. He has to tread a delicate line between industry and conservation, between nature and mankind, between demands and desires – all within the conflicts that are created by the very nature of the Dales rocks.

Villages of Stone

Characteristic dated lintel over a cottage front door in Kettlewell.

Old stone buildings, in the villages and on the farms, are one of the delights of the Yorkshire Dales. If there is a typical traditional house in the Dales, it is a long two-storey house built of grit, limestone and flagstone. Its walls are irregular blocks of local limestone or grit, with strong corners of larger grit quoins, with lintels and window mullions of carved grit, and with a low pitch roof of local flagstone. A special feature of a Dales house is often a dated lintel over the front door. Carved into the grit are the initials of the couple for whom it was built, alongside the date that it was built; many are from the late 1600s, the time when stone replaced timber for most new buildings.

The oldest stone houses were made from poorly shaped stones, many gathered from the fields and river beds, though few of these are anything but ruined remnants today. The skills of masonry had long been practised in building the castles, abbeys and priories in and around the Dales, but only in the 1600s was more care taken in building much better houses and barns.

In the typical Dales house, the main walls were built of local stone, simply to reduce the effort and costs of transporting heavy stone. It often came from the nearest rock outcrop, just working back into a low scar or crag to leave a face a little higher and steeper than its original, and soon blending back into the landscape. Grit (or sandstone) was always favoured, because it can be worked more easily than can limestone to produce roughly squared blocks that fit better into solid walls. Some of the sandstones split easily along bedding planes; the blocks they yielded may have been smaller but they were assembled into a wall almost as easily as bricks. Limestone rarely breaks along parallel bedding planes, so building with it took more skill, and it is only widely seen in the southern Dales cut wholly into the Great Scar Limestone.

Traditional stone houses in Arncliffe. On the left, built in 1747, with walls of rough limestone and grit, lintels of grit and a roof of flags. On the right, built over one hundred years later, with its front wall of dressed grit, quoins, lintels and mullions of sawn grit, and a roof of slate.

A cottage in traditional style, 'Dales vernacular', in Hawkswick village, Littondale.

Castle Bolton. But many other Millstone Grits and Yoredale sandstones also yielded very usable stone.

Flagstones, the thinly bedded sandstones, were mainly supplied from quarries and mines in the Yoredale beds of Wensleydale. Flags ideally about 50mm (2in) thick were used to pave the ground floors in houses and also the paths and farmyards outside. Large flags also made excellent 'throughs' that linked both faces of a thick house wall, and levels of them are a distinctive feature of the houses in some parts of Wensleydale. Lintels for the houses in Dentdale were made from even thicker flags taken from quarries by Scotcher Gill, in the north side of the dale, and at High Pike, on the saddle from Deepdale over into Kingsdale. In special demand were flags more like 25mm (about an inch) thick that could be used for roofing. These thin flags are known in the

Key elements of the house structure were then made with dressed stone – material that was supplied in blocks accurately shaped by hammer and chisel in the days before effective rock saws were available. Stone was roughly dressed for quoins – the larger blocks stacked up each corner of a house to improve strength and stability; stone was more finely finished for the mullions and lintels – the sides and tops of openings for windows and doors. Dressed grit was valuable enough to be transported further from the best sources. In Wensleydale, the best large quoins came from quarries in the Ten Fathom Grit across the Greets Hill slopes above Apedale, north of

Elements of a traditional Dales house – rough stone walls mainly of limestone, grit quoins and frames, and a flag roof, on a cottage in Stainforth.

Thick and rather heavy flags are used to re-roof a barn on a wet day in Wharfedale.

Dales as slates, or sometimes as stone slates, and were often supplied from the same quarries that produced the thicker flags, just by selection of the beds that parted more thinly.

Real slate has only been produced in the Dales by the various little quarries in the two glens above Ingleton, where it forms the finer-grained beds between units of coarser greywacke. Slate is much stronger than flag, and can also be split down much more thinly, so it would make a roof both more durable and much lighter in weight. It was only used widely around Ingleton, and was later supplanted by much better Welsh slate, which was available when the railways arrived.

Good Welsh slate also now lies over various Dales houses that were re-roofed, after timbers had sagged under the excessive weight of local flagstone. This trend towards imported material has been driven by practicality, and has somewhat diluted the Dales vernacular architecture. Dressed grit from demolished mills in Yorkshire towns is often now brought in for renovations and extensions on houses in Dales villages. But flagstone is the greatest shortage, and is always needed for repair and renovation where the architectural character is well worth conservation in the villages. Much of the roofing flag now comes from India; fortunately, this material is almost indistinguishable from local Yoredale flagstone. In similar vein, slate for repair work now comes largely from Spain; this is not as good as Welsh slate but is cheaper, and it does look a bit more like the Ingleton slate that is no longer available.

Times change, and choices of building stones evolve. But brick remains a rarity in the Dales, and most of the local architecture retains a character unchanged for centuries. Though the houses are built by man, their natural materials reflect so well the geological character of the land – and, save for the cliffs of white limestone, so much of this lies hidden beneath pasture and meadow. It is good that villages of beautiful stone houses remain an essential component of the Dales landscapes.

Stone houses in Gayle village, with a projecting course of flagstone that is a hallmark of Wensleydale.

continued from p.174

of today's much larger quarries directly employs only twenty or thirty men, but also gives work to around fifty truck drivers and various peripheral trades.

Some quarries are unwelcome scars on the Dales landscapes, and visitors wonder how they ever gained planning permission. But permits were only applicable after 1946, and quarries already active at that time were granted permits automatically in an expansive atmosphere of post-war re-development. Since then, some permits have been relinquished in exchange for permission for more suitable expansions elsewhere, and new permits are now scarce indeed, especially within the National Park. But the demand for stone is relentless.

Within the Dales, limestone is, and long has been, a major resource. Almost equal in demand is the grit, whether it is greywacke, gritstone, sandstone or flagstone. Slate, chert and 'marble' have been extracted on smaller scales in the past. Each Dales quarry, new or old, is based on one of these stone resources.

Limestone and marble

As early as the twelfth century, large quantities of lime were produced within the Dales. Limekilns became a feature of the landscape, each beside a small limestone quarry, many of which are now barely distinguishable from natural scars. The lime was first needed to make mortar, but demand for agricultural lime increased notably in the sixteenth century. By 1602, the lime industry was so intrusive that new kilns were even banned in Giggleswick, but a renewed boom in liming to improve fell land came with the Enclosure Acts around 1800. Demand for agricultural lime soared again in the 1930s, but has declined since, notably with the more recent ending of grants for land improvement. Today, most of the strong limestone goes into concrete aggregate; though much of the Great Scar Limestone is 98.8 per cent pure calcium carbonate, demands for the steel furnaces and

A field kiln in Chapel-le-Dale that once produced agricultural lime from local stone burned with local peat and wood from nearby trees.

Inside the line of chambers in the preserved Hoffman kiln at Langcliffe Quarry.

the chemical industry are now minor within the Dales.

Hundreds of stone-built limekilns are scattered across the fields of the Dales, though all are now disused and in various states of dereliction. They stood wherever limestone was available, and commonly beside a peat hag or an outcrop of thin coal in the Yoredales. Digging out the limestone to feed them left small scars hardly worthy of being called quarries, and most of these are difficult to recognize since they have weathered into the landscape. The kilns burned the stone to make lime, by taking the calcium carbonate down to the oxide; and this was often left in the rain to hydrate into slaked lime. Each kiln was worked by throwing in a mix of limestone and fuel, and drawing lump lime out from the base.

Quarries became larger to feed the next generation of industrial kilns, including the Hoffman kilns that allowed continuous production without the cooling delays inherent to single kilns. The kiln of this type at the Langcliffe Quarry had twenty-two chambers where surplus heat was transferred from the burning chambers to pre-heat the next, so that the fires took six weeks to make a circuit through all the chambers. Built in 1872, Langcliffe's Hoffman kiln is preserved as a rather fine but little-known site of industrial archaeology, below its quarry that now appears as the broken cliffs on the southern end of Stainforth Scar. On the opposite side of Ribblesdale, Foredale Quarry produced lime until 1958, working the massive, white, pure Cove Limestone, but taking little of the underlying, dark and muddy Kilnsey Limestone.

The enormous Horton Quarry was started in 1888 by an enterprising Irishman, who saw an opportunity for a new lime works when the railway was built up Ribblesdale. A huge extension was permitted in 1940s, in the heady days post-war, when its pure limestone was needed for a renewed steel industry. Since then, needs have changed, and its output is now entirely for concrete aggregate. It is sad that the rail access was taken out in the Beeching years, because the quarry's road traffic has a major impact on Settle and the villages along southbound roads; a renewed rail link will

A massive processing plant stands across the entrance to the Swinden Quarry, which is steadily removing the entire inside of the limestone reef knoll.

greatly improve the situation. This should come when the limestone quarry expands downwards into the Ingletonian greywacke under its present floor and combined annual output of limestone and greywacke will reach towards a million tonnes. The intervening 8m (25ft) of muddy Kilnsey Limestone will have to be taken out and sold as low-value fill, but the quarry operators have been fortunate that no caves (and no cave sediments that would be so much waste to be removed) have been met by the advancing faces. The quarry has notably little water inflow, as it seems that drainage within the limestone is away from the quarry or down into the Kilnsey Limestone beneath the current workings.

Most of the Dales limestone lies within the National Park, but most of the quarries in the Park have now ceased operations. The older quarries, like Meal Bank, Storrs Common and Ribblehead are starting to blend back into the landscapes, forming cliffs and crags that mimic many of the natural scars except for their locations out of context with the local geomorphology. Still looking like quarries are the more recent closures, at Cool Scar, Skyrethornes and Giggleswick, where consents have run out or have been traded for other permissions. Swinden Quarry, working its way through an entire reef knoll near Grassington, is the only quarry within the Park currently producing over a million tonnes of stone per year.

Just outside the National Park, production from the Coldstones Quarry in the Greenhow anticline almost matches that from Swinden. Just outside Skipton, the Haw Bank and Skibeden Quarries coalesced along the anticlinal ridge of dark, basinal limestones before they closed in 1994; they were the largest of various quarries along the northern edge of the Craven Basin. Also away from the Great Scar Limestone, the Main Limestone, above the Yoredales, has been worked since the 1850s by a handful of quarries along its outcrop from Leyburn to Redmire Scar, on the north side of Wensleydale; two still supply both concrete aggregate and also flux to the steel industry on Teesside.

Away from big industry, smaller quarries all across the Dales have been opened to supply building stone for the farms and villages. But limestone was never favoured because it cannot be shaped or dressed as easily as grit. Demand has long gone and most of the little quarries are barely distinguishable from

Stone Viaducts

Largest of all the stone-built structures in the Dales are the railway viaducts. The finest of these are along the scenically spectacular Settle–Carlisle line. This dramatic line was built by the Midland Railway between 1864 and 1875, simply because they wanted a route to Scotland over their own tracks. It was a great piece of engineering, and a shanty-town for 2000 workmen temporarily transformed the landscape at Ribblehead. The whole railway was threatened with closure in 1984, but was saved in 1989 after a campaign revealed dirty tricks and fallacies generated by British Rail. One claim was of excessive costs to maintain the viaducts, but after modest repair works, at about half the proposed costs, they still stand today.

Ribblehead Viaduct is one of the Dales' great landmarks, striding across Batty Moss where the glaciated trough of Chapel-le-Dale leaves that of Ribblesdale. Over 400m (1300ft) long and 32m (105ft) high, its massive piers of dark limestone are capped by twenty-four brick-built arches. Each pier is founded on Great Scar Limestone, where up to 8m (26ft) of peat had to be removed to establish the bedrock footings. To build the piers, blocks that weigh up to 8 tonnes were worked from the Hardraw Scar Limestone in trackside quarries in Little Dale, just to the north. Between the massive blocks that face the piers, a poorly placed fill of rubble lacked the strength that dry stone wallers had long achieved by building a core with carefully placed smaller stones. A bitumen seal was placed over this, just below the trackbed, but was carelessly installed around timbers that subsequently rotted away. The resultant holes let in the water, and the saturated core of the piers then became serious weaknesses. A new membrane seal and extensive grouting of the core have effectively stabilized the piers – whose facing blocks of limestone are still very sound.

Upper Dentdale has two more splendid viaducts, Dent Head at the top end, and Artengill across a large tributary valley. Each is about 200m (650ft) long, and Artengill is just the taller at 37m (120ft). Both were built

Ribblehead Viaduct strides across Batty Moss, while patches of snow remain on the slopes of Whernside.

Artengill Viaduct stands in front of a flat-topped heap of rock excavated from the adjacent cutting in the slopes of Dentdale.

of blocks of dark Simonstone Limestone, extracted from small quarries close to each viaduct; this is the same rock bed as the more fossiliferous stone that has been worked locally and polished up to become Dent Marble.

The fourth big viaduct on the Settle–Carlisle line is Dandrymire, which crosses the head of Garsdale. Low and squat, this lacks the elegance of its neighbours, and was never part of the initial railway design. The original plan was to place the railway on a low embankment, but Dandry Mire is a peat bog of extremely soft ground. The embankment was built out from the north side by end-tipping rock rubble that had been won from the cuttings and tunnels. This was intended to compress the peat and turn it into a load-bearing soil, but the peat was so soft that it was just squeezed out the sides while huge amounts of unstable rubble sank into the bog. The remedy was to continue with a viaduct founded in excavations down to stable bedrock of Middle Limestone; the piers were built with blocks of strong, fine-grained sandstone extracted from nearby Yoredale outcrops.

Viaducts are normally built with the nearest available strong stone, even if this means specially opening new small quarries, because transporting stone onto site can be unduly expensive. So the three big viaducts on the Settle–Carlisle line are all made of Yoredale limestone. But others are different. The abandoned viaduct that is still such a feature in Ingleton village was made of Carboniferous sandstone brought in from near Bentham, while smaller structures in Ribblesdale have utilized the local Horton Flags. The engineers of old knew well to relate to the Dales geology, and their structures therefore live on as components of the grand landscape and mirrors of the bedrock.

Blocks of local Simonstone Limestone were used to build Artengill Viaduct in Dentdale.

natural scars. In times gone by, an extra demand for limestone was for ornamental use, where oddly shaped and curved stones were favoured to top out garden walls in rather garish style. This stone was taken from the top bed, that most carved by dissolution, on a limestone pavement, and was therefore highly destructive of a precious natural landform; its production is now banned by well-enforced limestone protection legislation. Pavements on Asby Scar and around Morecambe Bay suffered the most, but some pavements within the Dales, including patches on Ingleborough, are missing some of their more deeply carved clint blocks.

Any limestone that can be cut and polished is known in the trade as a marble, and the Dales had a few that earned this title. Dent Marble may be black or much paler, and some is made distinctive and truly spectacular by its abundant large crinoid stems. For almost the whole of the nineteenth century, Dent Marble was worked from the Yoredale sequences high on the slopes of Dentdale, notably around Scotcher Gill, and also in Garsdale. Best known is the beautiful stone, grey or cream and packed with crinoids, that came from the Binks Quarry in the Underset Limestone high on Great Coum. This was the Dent Marble used to make the altar steps in Dent church, where it is laid alongside blocks of black marble with few fossils that was worked from the Hardrow Scar and Simonstone Limestones. A comparable black crinoidal limestone was known as the Nidderdale Marble, extracted in Blayshaw Gill, near Lofthouse, in the Middle Limestone; a small quarry by the Gill was worked from 1869 until the 1920s, and may have been the source of decorative stone used in building Fountains Abbey, more than 500 years previously.

Gritstone and greywacke

Though grits form half the hills between the Yorkshire Dales, they are not the target of

Naturally weathered clints form a slightly grotesque ornamentation on top of a garden wall in Ingleton.

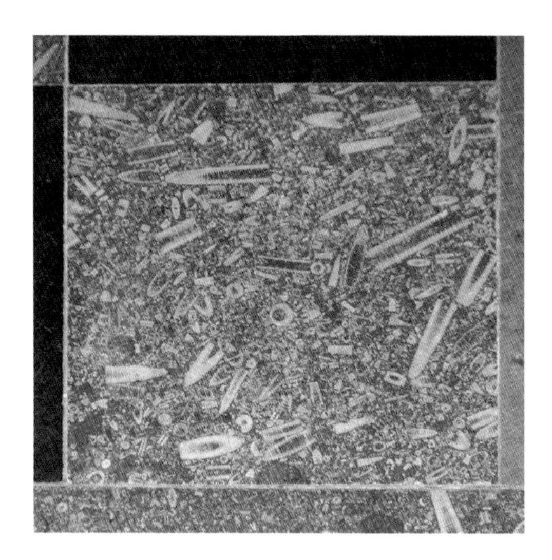

Polished slabs of Dent Marble, packed with crinoid fossils, form the altar steps in Dent church.

today's quarries. The hard stone in demand is the greywacke that lies within the basement underneath the Carboniferous Limestone. Known in the aggregate trade as gritstone, greywacke is the first choice for roadstone that goes into the surfacing of motorways and main roads – because it is very strong and it does not polish. Its sand grains give it a permanently rough surface even as it wears down; it's like trying to polish sandpaper – compare that to granite or limestone in kitchen worktops. So its 'polishing stone value' (PSV) is high, and it has excellent skid-resistance. This is why it passes the tests of rarity and value that allow it to continue to be quarried within a National Park.

The best greywacke for roadstone is the coarser material, and any interbedded fine-grained slate is unusable waste. The Ingletonian rocks are the best, and the Ingleton Quarry produces nearly half a million tonnes per year from a tight syncline of the best greywacke. It is all used as the wearing course on roads, by feeding eleven tar-coating plants in Yorkshire

Steep and vertical greywacke is followed into the depths at the Ingleton Quarry.

and Lancashire. But the quarry is trapped between the River Doe and the Chapel-le-Dale road, so it will run out of stone in about 2012, when production will switch to a hole in the floor of the existing limestone quarry at Horton. The Ingleton Quarry will fill with water, to become a lake 120m (400ft) deep, with an overflow cascade into the Doe near Snow Falls. Just up the valley, the Old Granite Quarry takes its incorrect name from a band of very coarse greywackes containing pebbles of pink feldspar, but reworking this is hardly favoured when it is so prominent in the Ingleborough landscape.

The Silurian greywackes in the floor of Ribblesdale are rather fine-grained but their aggregate does provide a polishing resistance high enough for their use as roadstone. The Helwith Bridge and Dry Rigg Quarries are both in the Horton Flags. The former is already a deep lake, and Dry Rigg will close before 2010, when waste heaps will be used to backfill part of the quarry, so that the adjacent and treasured wetland of Swarth Moor can spread back across it, except for a deep lake under the face at the western end. Arcow Wood Quarry extracts the turbidite greywackes from the Austwick Formation, and will cease working at about the same time.

Grit and sandstone

Though they are topographically quite resistant, the various grit beds in the Carboniferous are not tough enough to be used as aggregate. The rock abrades too easily when individual sand grains are scraped off exposed surfaces. But for the same reason it is easily sawn and, unlike limestone, can be easily worked (or dressed) into blocks that make excellent building stone. Most is locally described as sandstone, but coarser material may be known as grit. The stone is a key feature in most old Dales buildings (*see* page 178), and the remains of a little old quarry can be found on the nearest outcrop to many villages and farms.

Richmond Castle was built around 1160 with Yoredale sandstone that came from Gatherley Moor, north of the town, and between 1650 and 1700, many buildings along Skipton's main street were built with Millstone Grits from quarries long abandoned on Embsay Moor. Burnsall has one of the loveliest of the many Dales bridges that were built in the 1600s and mostly widened in later years, all with local stone from the Millstone Grits or the Yoredales. Finest of all the building stone came from the Scotgate Ash Quarries, which were worked in the fine-grained, micaceous Libishaw Sandstone just above Pateley Bridge. Between 1860 and the 1920s, these quarries produced around 10,000 tonnes of stone per year that was used in buildings all over the country. Various other quarries worked the same sandstone in Nidderdale, including one

High above Pateley Bridge, an old quarry in Libishaw Sandstone is now just a dumping site.

Hilltop Quarry, still extracting the Tan Hill Grit at the top end of Swaledale.

opened to build the dam at Gouthwaite. Scar House Dam was built thirty years later with a concrete core faced with Red Scar Grit won from a quarry specially opened on Carle Fell, just north of the dam. Almost directly opposite, on the south side, another quarry was later opened in the same grit for stone to face the Angram Dam further up the valley.

Only one quarry still operates today in the Dales sandstones. At the top end of Swaledale, Hilltop Quarry takes about 400 tonnes per year from the Tan Hill Grit. This is a good strong sandstone, which is now nearly all sawn into paving slabs. Most of these show concentric banding of red, brown and orange colours on their sawn faces; known as Liesegang rings, these are created by different minerals of hydrated iron oxides left by water moving very slowly through the rock's pore spaces. The same quarry was once famed for its excellent flagstones, but most of these were worked out long ago.

Flagstone and slate

Flagstone is a strong sandstone that can be 'riven' or split along natural bedding planes, into sheets about 50mm (2in) thick. Sawn, guillotined or chiselled into square slabs, these make durable paving flags. Many Dales farmyards were paved with Yoredale flags more than 200 years ago and are still in excellent condition. Old flag quarries trace the outcrops of the best beds, most notably along both flanks of Wensleydale. A particularly good bed of flags, nearly 2m (6ft) thick, lies just below the Middle Limestone, and was worked in quarries and mines above the villages of Sedbusk and Burtersett, with a peak of activity in the 1890s. In the horizontal beds, level tunnels were driven up to 400m (1300ft) into the hills, with pillar-and-stall workings on each side. At an active face, the men cut out a soft bed above the good stone until they reached a major joint where wedges could be hammered down. At the same time wedges were used to split the beds, so that slabs could be pulled off one at a time working downwards. The same bed was worked in mines above Carperby, where it reached over 4m (13ft) thick, so the mines had very spacious galleries with advancing faces nearly 10m (30ft) wide. Stone was split into flags at the face, and the roof behind supported by

The spacious gallery in the Carperby flagstone mine, where waste material forms a massive pillar to support the roof behind the advancing face.

massive pillars built out of a large proportion of waste material. Sadly, the thick flags that were extracted were not the most durable, and production ceased in the 1930s.

From the many mines and quarries in both Wensleydale and the other northern Dales, the best flags were used for paving, poorer material went out as wall stone, and the waste was abandoned, some of it underground in areas already worked out. Thinner flags were used for roofing, and were therefore known locally as slates. Many of these were produced by taking thicker blocks from the mine and leaving them outside through a winter; frost would get into the stone's pore-water and start to heave the rock apart along its natural bedding weaknesses – making it much easier to rive or split into good thin roofing material the next summer.

The only true slates in the Dales come from the metamorphosed Ingletonian rocks of the basement, where they form the finer grained material between the coarser greywackes. They do not have perfectly developed cleavage, so they could not easily be split into roofing slates of the highest quality, but they were still often thinner and therefore lighter than the thin flagstones. A series of small quarries are still recognizable in the Ingleton glens (*see* page 90), though they all ceased working soon after the railways could bring the better slate from North Wales.

Not to be forgotten is one flagstone that is completely separate from those in the Yoredales. Around Helwith Bridge, in Ribblesdale, the Horton Flags were long worked from a number of quarries, including the one that is now an anglers' lake. Thick flags of the strong, dark blue stone have been used to build most of the houses and walls in and around the small community, and this rock is notable for some of the very large flags that can still be seen at Horton church and elsewhere. Though Horton church was built by the monks from

Jervaulx nearly 800 years ago, its flag floor was a much later addition.

Millstones and grindstones

It would only be right that millstones were made from some of the Millstone Grits in the Dales. Between the floor of Nidderdale and the crags of Brimham Rocks, the Millstone Quarry produced millstones up to 2m (7ft) in diameter from a grit bed below the Brimham Grit, but the industry was never large in this part of the Pennines.

Very hard stones were needed in the pottery industries to grind down clays in the days before hardened steel was available. Chert was an ideal material but was not easily obtained in blocks large enough to use. Some was extracted first above Fremington Edge, east of Arkengarthdale, but it was then quarried and mined on a larger scale between 1922 and 1950 at the Hungry Chert Quarries beside the Hungry Hushes high on the western side of the same dale. The miners could not afford to break the chert by using explosives, so they exploited natural joints to work individual beds that yielded intact slabs up to 400mm (1.3ft) thick. Chert mining is one of the least known of the Dales' bygone industries, but it has added yet another layer to the great diversity of the Dales geology.

Large chunks of Horton Flags form the footway and the roof to the lych-gate at Horton Church.

Large and thick flagstones lie abandoned outside the Carperby mine.

CHAPTER 12

Farming the Rock Landscape

Outside the areas and times of profitable mining, and beyond the scattered pockets of local industry, farming has been the big factor in people's lives within the Dales. Man is very clever at adapting to the local conditions imposed upon him by climate, terrain and geology, but the Yorkshire Dales have hardly given him an easy ride through the years. Slopes too steep to be put to the plough, and soils that are too thin or too wet, restrict most of the productive farming to the floors of the dales, but there are few areas even on the highest ground where farmers past and present have not made their contribution to the texture of the Dales landscape.

Man's early impact

High in the scars behind Settle, Victoria Cave has yielded bits of pottery, bronze brooches and bone combs that show man was already living in the Dales around 12,000 years ago, soon after the last retreat of the glaciers. Limestone caves never made very comfortable dwellings in the Dales, but fire hearths, charred animal bones and crude tools show

Early farmers cut lynchets into the slopes of colluvium and moraine to improve land value in Littondale.

Cave Ha, a rather open rock shelter in Giggleswick Scar.

that a handful of open entrances and rock shelters were so used. Other caves became burial sites, including the little chamber near the top of the Elbolton Hill reef knoll, in Wharfedale, in which four out of a dozen Stone Age skeletons were found in an upright sitting position.

These first Dales dwellers had to develop the land in order to survive, and their descendants were soon playing an increasing role in shaping the landscapes that are seen today. Around 5000 years ago, so at about 3000BC, the Neolithic (Late Stone Age) population increased in numbers and they started the major clearances of the forests that originally clad all but the highest land. This took them into a period of warmer climates, which lasted until a return to cooler and wetter conditions around 1000BC. At this time, in the Romano-British age, many Dales people did still occupy some of the more hospitable caves, including the spacious Cave Ha and other south-facing rock shelters along Giggleswick Scar. But many more were living in huts of stone or timber built among their fields, especially on the limestone areas that had already been cleared to create open grassland.

Increased tendencies for cropping on arable land heralded the ancient lynchets – level or gently inclined field terraces cut into slopes in order to create better land for ploughing. Many lie on the ramps of thick soil and colluvium along the foot of much steeper dale-side cliffs. Some date from the Celtic and Anglian farmers of the sixth and seventh centuries, while more were created five hundred years later under the pressures of medieval farming. Only further declines in climate have seen the end of cropping on them, so that the ancient lynchets are now just grassy banks in sheep meadows.

Global cooling continued right through to the Neoglacial, the Little Ice Age in Alpine regions, of the sixteenth century. It changed the farming in the Dales, first away from arable cropping, and then away from pigs and cattle in favour of sheep. From soon after 1066, until their dissolution around 1538, outlying monasteries managed huge land holdings in the Dales, and these were dominated by sheep farming. Cattle farming, for both beef and milking, does continue, but only on the floors of the dales. More than a dozen herds, each of over a hundred head, are held in farms along lower Wensleydale, but even there they are kept inside for half the year. In summer they graze pasture on the lower valley sides, while meadows alongside the river yield two or three cuts of good grass that goes to silage and the winter feed. Meadows are smaller in the upper dale, so cattle numbers have steadily declined. Sheep have taken over, and remain a potent force within the Dales' landscapes, especially on the high fells.

Enclosing the land

As farming practices steadily involved, they

Dry stone walls around West Burton, in Wensleydale, show the patterns of three successive stages of increasing size of enclosure: first, around the village, clustered round its splendid green; second, along the dale floor, mostly below the 200m (650ft) contour, and with the isolated byres; and third onto the high fells of Penhill, above the 300m (1000ft) contour that lies along the crest of a low scar formed by the Middle Limestone.

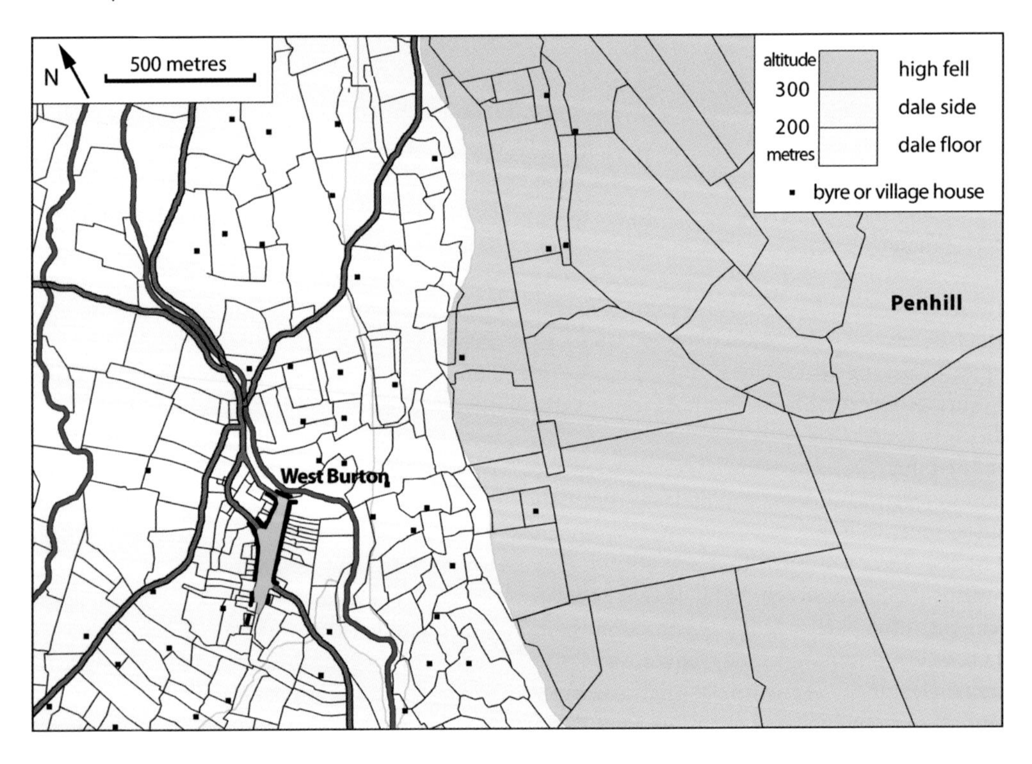

A dry stone wall heads up over Fountains Fell in order to follow the boundary of an upland enclosure.

responded to the geological conditions that controlled the natural landscapes. Except where bare pavements were beyond any scope for development, the limestone fells between the southern dales were grazed by sheep, to become the wide-open grasslands that are still a hallmark of the region. Farther north, the great grit moors declined almost into wilderness, too high, too wet, and too bleak for any serious farming. In between, the varied Yoredale rocks create easier slopes where farms could develop on usable terraces near supplies of clean spring water.

Bedrock is not always the key factor, especially where it is masked by thick blankets of glacial till, as this commonly creates heavy wet soils that are of little value even where they overlie well-drained limestone. The dale floors have offered most of the best land, with some good soils on areas of alluvium; in front of Kilnsey Crag, the flat floor of Wharfedale is a reach of ancient lake sediments that support lush meadows. The floor of Kingsdale is a gently sloping alluvial fan, but this was flooded so frequently that an artificial stream channel was excavated around 1900. This protects the fields belonging to Braida Garth, the farmhouse that stands up on a lateral moraine safe from any flooding. Farms and villages have always avoided the wetlands, and many lynchets also reaped the benefits of sites just up onto the dale-side slopes.

Farmland in the Dales falls into three distinct types: the tiny proportion of rich valley-floor meadow, the larger extent of permanent pasture that spreads up the dale sides, and the huge extents of upland moor. The farming value of each is clearly very different, and is reflected in the three phases in which the land was enclosed – by stone walls. These are almost indestructible, so they remain as major components of the modern landscape – and there are more than 8000km (5000 miles) of dry stone walls within the Dales. Control of straying animals prompted the building of the earliest walls, from about 1300 onwards, and these took crooked lines to enclose small fields around the villages. Mainly during the eighteenth century, systematic enclosure of larger fields spread the walls across practically all the dale floors. Then the iniquitous Enclosure Acts, from about 1770 onwards, enabled well connected landlords to grab large slices of the upland moors and fells by building those long straight walls that rise from dale floor to reach right over the bleakest summits.

Scattered between the boundary walls, there are two more stone-built features that typify the Dales – byres and bields. The byres are field barns that house cattle through the Pennine winters. Whereas many traditional farms were stone long houses, with family and animals in adjacent sections, the byres had the benefit of being at the food source. They were built in the meadows so that cut grass could be placed directly in the upper floor, ready to drop down to the cattle when they came in from the winter. The floor of Swaledale has the best of the byres, with one in every meadow, but they stand throughout the dales, a few still in use, though most falling into disrepair. The bields are also shelters, but are just short lengths of wall behind which sheep could retreat from the winter winds. A number of them still offer shelter on exposed parts of the high fells, but others have collapsed into isolated piles of stones that seem out of place on a heather moor until their origins at the hand of man are remembered.

Walls across the landscape

Dry stone walls are now a signature of the Dales landscape. Though most are old, they are still being repaired, and some are built anew. They keep a handful of artisans in steady work (*see* page 198), and walling competitions at the annual country shows are just a bit of whimsy within a serious craft.

A Dales field wall has no mortar, and needs none when it is built to designs and practices that have matured through centuries of experience. Its base is half as wide as its height, to ensure long-term stability, with faces that lean in at a good batter, so that some modest settlement and rotation will tip neither past the vertical. Each face is a carefully built stack of stone blocks whose good fit together is the mark of the waller's skill. The core is a laid mosaic of smaller stones, except for the largest blocks that are placed as 'throughs' so that they are part of both faces, linking them together through the core. A local variation is traditional in Wensleydale, where lines of 'throughs' look good but are not so strong because their tops are planes of weakness

Built to winter cattle, but now used for sheep feed, a stone byre stands amid meadows on the floor of Littondale.

ABOVE: Field barns and dry stone walls texture the meadows on the floor of Littondale, just above Arncliffe.

BELOW: Dry stone walls at their best, along the floor of Wharfedale.

Dales Waller

John Heslegrave loves working with natural stone. He lives in Leyburn, and traditional dry stone walling is his forte.

Unlike many a farmer who has to rebuild his field walls when they suffer the ravages of time, John started drystone walling as a hobby, while pursuing a long career in business. But the outdoor life has an undeniable appeal, so he evolved into a full-time waller some years ago, and his son Alan now works with him to make it a true family concern. While half their time is spent on building stone walls and features in domestic gardens, the other half takes them out on the fells, to work on the dry stone walls that are such an important part of the Dales landscape.

Most of their work on field walls is repair and restoration. John reckons to build up to 4m of full-height wall in a day; any more is rushing the job. That is well in line with his favourite phrase from a respected Dales waller – 'a good waller makes five yards a day, while a bad waller makes eight'. Extra time is needed for a wall repair, because it usually starts with clearing the failed old wall right into the soil in order to re-create a stable and level footing. Then it's down to patient and skilful selection of each stone to build strong and tight faces on both sides of the wall. At judicious intervals, the larger stones are used as 'throughs', and the smaller stones are placed in the core of the wall with equal care – tipping in of loose fill is totally unacceptable. It's heavy work; building 4m (13ft) of wall means lifting and placing over 5 tonnes of stone.

His eyes constantly scan over the pile of stones to spot the individual of the right size and shape; dressing a stone with hammer and chisel is rarely needed, and is mostly to achieve a close fit inside the wall. A key role of the walls is to shelter sheep from the worst of the Dales weather, but an ungrateful sheep will use projecting stones as footholds to reach greener pasture or an attractive ram on the other side. Respecting the needs of the farmer with heavy investment in his stock, the waller endeavours to keep the faces tight.

John does have a stone saw, but uses it mainly to cut sandstone accurately and quickly to fit into the geometry of domestic projects. He does believe that the old

Alan looks down at a wall that John and he are rebuilding in Wensleydale; the section beyond, tilted over by soil creep, is also due for serious repair.

Dry stone walls that have stood for hundreds of years.

wallers would have used saws had they been available hundreds of years ago. A saw now saves time and effort when large or brittle stones can be cut to shape for a wall-head or to make a pair of good 'throughs' from an oversize slab. And a very large block can be broken by the traditional method of hammering wedges into a shallow saw-cut. Some of the heaviest stones are held back to sit as large 'copes' along the finished crest; in John's view, these are the best defence against a wall's two worst enemies: sheep who will climb over any but the highest wall, and cattle who will nudge smaller cap stones off the top.

The waller repairs and rebuilds field walls with whatever material is already on site, whether it is limestone, grit, finer-grained sandstone or even flagstone. Some well-bedded and closely-jointed limestones from the Yoredale sequences can yield very square blocks; these are easy to work with, but they create a wall so uniform that it loses the character of a rugged appearance. An old wall is largely its own supply, but additional matching stone may be brought in from the big quarries or from old abandoned structures.

John and Alan enjoy their work on domestic projects, where each job is unique to its design and situation. John does admit to a modest yearning – he would love to be asked to build a dry-stone bridge. But the routine of working on field walls that can be long and uniform is for John overshadowed by the simple pleasure of being out on the open fells, even in the less clement of Dales weather. To work on the field walls is to become a part of the countryside and its long agricultural and social heritage. The walls that John builds become a part of man's subtle imprint on the natural terrain; and that partnership of man and nature adds so much to the special character of the Dales landscape.

within the wall. Copes also give strength along the crest by straddling the two faces, as do the wall-heads where facing goes across the end of a wall, either into a gateway or at an owner's boundaries.

Built to about 1.5m (5ft) high, walls have up to 1.5 tonnes of stone per metre length. They are heavy structures, but most are founded within the soil. How deep is down to the waller's judgement, but a base too shallow can let soft soil be squeezed out by the weight of stone, and this almost always happens unevenly so that rotation soon leads to failure. Next to attempts at climbing over – by either hungry sheep or careless ramblers – the walls' greatest enemy is soil creep. The impact is often recognizable where a once-straight wall now waves and winds across a slope. A wall that is curved even where it is founded on a deep base indicates that soil creep can be active to a depth of a metre or more, most notably on slopes of clay-rich debris in the Yoredale country.

The stone with which a wall is built is a reflection of the local geology, either of the bedrock or of coarse debris within the soils. Many of the earliest family-built walls were built with clearance stone – material that was pulled from the fields in order to improve the soil before ploughing could be started. Others were made with river cobbles. Both these sources yielded poor, rounded and mixed stone, so that the walls may not have been the most durable. For better material, dozens of small local quarries were opened, especially during the heydays of walling around 1800. Limestone in the southern Dales, sandstone in the north, and flagstone in Ribblesdale were all used. Except for the flagstone, most block shapes were still irregular, but at least they were not rounded. Most of these little quarries are now barely distinguishable from natural

In Gunnerside Gill, the green grass of lime-improved enclosures contrast the brown bracken that has spread over the open fell.

A farmer near Buckden has to feed his sheep on the fell above Wharfedale after snow has buried their feed grass.

scars. Some lasted till later as they were taken over by the local councils, but all are now abandoned. New walling stone is today just a by-product from the big aggregate quarries.

Liming the intakes

Intakes were parcels of land that were walled and improved distant from the farms. They lie scattered across the poorer land on steeper dale sides and out on the high fells. Some belonged to valley farmers, but many of these intakes were worked by the Dales miners. The underground labours of these men were usually paid for at piecework rates, so they could develop some seasonal flexibility and then benefit from spells of work in the fresh open air on their own moorland intakes. Once walled, the intake land was improved, primarily by liming. Poor, heavy, wet soils benefit from the addition of lime, because it flocculates the clays and so aerates the ground, allowing both oxygen and bacteria to work to good effect. Lime also counters the acid effects of nitrates, and allows more fertilizer to be applied; it makes the soil sweet, or less sour. A maintained intake is recognizable as a swathe of good fescue grasses in place of the reeds of wetter, acidic ground that may still lie just outside its walls. Many abandoned intakes have only broken boundary walls and encroaching reed grasses, though their names may still survive on detailed maps.

Agricultural liming brought large areas of new land into cultivation, both on the higher parts of the dales' floors and in the scattered moorland intakes. It was big business in the

Improved pastures surround a hill farm in Sleddale, but give way to the open fell above.

seventeenth and eighteenth centuries, and it had a major impact on the Dales landscapes. Most of all, there is a significant visual contrast between the sweet green fields on limed soils and the brown fells of poorer land that surround them. But there was also the impact of forest clearance to supply the wood needed as fuel in many of the limekilns. Rather less conspicuous were the scatter of small limestone quarries to feed the kilns and also the various small coal mines, especially in the Yoredale country, that eventually supplanted wood to fire the kilns.

Land improvement on the poor soils that are so widespread on high ground between the dales could not rely solely on liming. It required good drainage and then follow-up with appropriate fertilizers. Another key was heavy grazing, to keep down new shrubs and trees while good grass could thicken. Combine all this with even wider clearances of woodland, and there are clear comparisons to be made with today's deforestation in the Amazon basin. Farming can change a landscape forever, and the Dales farmers of yesteryear certainly changed the Dales landscape from the purely natural to one that is blended with the imprint of man.

High fell, high moor

If farmers fashioned the Dales, then sheep fashioned the fells. Their unceasing nibbling takes the buds and shoots so that most shrubs and plants cannot get established. But grass grows from below, and it thrives with sheep grazing. Hence the open grassland, often described as sheepwalk, that is such a delightful feature of the Dales panoramas.

Hill farming equals sheep grazing, and is the big feature on land that is high and wet. In the Dales, this is land above the 400m (1300ft) contour washed by over 1250mm (50in) of annual rainfall. Sheep are bred hardy to survive the Dales winters, and the Swaledale is the classic breed; not for nothing is a Swaledale tup (a mature ram) the emblem of the Yorkshire Dales National Park. These are the sheep that are hefted onto the open hills. This is an inherited characteristic which determines that each sheep stays on, or returns to, her own heaf or heft, her own section of open fell, and will not stray across a broken boundary wall. It's an invaluable attribute when a farmer has to find his flock after a storm.

A workable hill farm has perhaps a thousand sheep, but is still only sustainable with support from government subsidies, which are now based on the areas of land. Each farm has access to more than 500ha (1200 acres) of fell for grazing. It also has a much smaller stretch of fields on the dale floor fields, and these are used for spring lambing in the nutritious new grass of meadows rich in flowers, herbs and clover. On alternate years, the meadows are cut to storage for winter feed. A snow storm brings work for the hill farmer, who has to carry food to sheep still out on the fells; they stay warm enough in their wool, but get very hungry when their food is lost under a blanket of white.

Geology has its own influence on the hill farming. Most of the limestone plateaus above the southern dales provide excellent high pasture with their good fescue grasses. But not all is perfect. Bare pavements have no meal appeal for sheep, and a blanket of glacial till on top of the limestone gains a cover of juncus reed grasses that have far less food value than the fescue. Also, sheep are notoriously stupid. Grass grows around the edges of potholes, so sheep fall down them. Every year, a few are lost when they plunge down deep shafts, while rather more are hauled back to the fell when they survive a slither down a less daunting pothole and are found by a searching shepherd or a visiting caver. The grit moors above the northern dales lack the potholes but also have much of the fescue grass replaced by heather. Though this can provide winter feed for sheep, it is not as nutritious. With the decline in sheep farming (largely induced by changes in farm subsidies), there is increasing economic benefit in land managed primarily for breeding grouse and pheasant. Whether the shooting is for sport or for food, the grouse moors are an increasing feature of the grit uplands between the dales. Huge swathes of purple heather create yet another component of the Dales landscape that is largely down to man's own actions.

Wharfedale sheep farm in winter.

CHAPTER 13

Landscape to Enjoy

More than 10,000 people live within the Yorkshire Dales. They include the farmers, those who make a living in the holiday industry, and also the many who can work from home (where a computer may be all that is needed, in place of the knitting skills of bygone centuries). Though the hill farmers especially face a lot of hard work, all who are lucky enough to live in the Dales extol the excellent quality of life. Parts of that quality are also enjoyed by the visitors, about eight million every year, of whom about half stay longer than a single day. To all these, the Dales offer fresh air, recreation and enjoyment in a most splendid part of the English countryside.

National Park

Designated back in 1954, the Yorkshire Dales forms one of the largest National Parks in Britain. Though based on the characteristic landscapes of limestone, grit and dale, the

Kettlewell is one of the finest Dales villages, on the floor of Wharfedale and overlooked by limestone crags.

Family campers at the foot of Gordale Scar on a holiday weekend.

Park boundary was drawn with some odd quirks. It excludes the very cavernous limestone of Ease Gill (because it is outside Yorkshire) and also Upper Nidderdale (because it was in the hands of Bradford Waterworks, though it is now designated as an Area of Outstanding Natural Beauty). Then it includes the Howgill Fells (because they are in Yorkshire) even though they are geologically distinct from the Pennines. But the National Park does constitute a recognizable entity in its fine upland terrains dissected by the long and deep glaciated troughs that are the dales themselves. Like the others in Britain, the National Park is essentially a planning authority, intended to serve both the people who live in it and also those who just visit. Such a role demands compromise, but conservation and leisure are given more respect while industry and economy are held back from dominance in planning and new development. Tourists, farmers and ramblers are just some of those who vie for space in the Dales, and all have their right to benefit from a piece of Britain's natural heritage.

PAGE 206: Exploring downwards: a caver sets off down a long rope beside the first big waterfall in Diccan Pot.

PAGE 207: Exploring upwards: climbers on parallel routes up the vertical limestone wall of Malham Cove.

It is a reflection of modern life that most visitors to the Dales travel by car. Furthermore many of these do not venture more than ten minutes' walk from the car park. A short stroll to a natural feature and a wander through one of the lovely villages may be the Dales experience beyond window-gazing at the splendid scenery. Perhaps fortunately, the Dales are laced by an uncommonly good network of minor roads, which are usually empty enough to attract cyclists. Geology accounts for the two worst cases of the quite modest road congestion – both the summer Sunday car traffic into Malham to reach its scenic wonders, and also the weekday lorry traffic south of the Horton quarries in Ribblesdale. Sadly, there is little available alternative to this

Horton Lane is a classic 'green track' up onto the Pen-y-ghent fells, popular with walkers and also still used in a modern version of herding sheep.

road traffic. The Settle to Carlisle railway offers great window-gazing, but delivers only handfuls of walkers to its remote moorland stations, and currently has no access from the quarries that lie beside it. And buses are a rarity in most of the Dales. So cars reign supreme, but they do bring plenty of visitors to enjoy the Dales away from the roads.

Recreation

For both young and old, 'outdoor activities' are growing steadily in popularity. A walk in the country always pulls in the largest numbers, but the Dales offer alternatives. Horse riding and mountain biking take enthusiasts across the highest fells. Sailing is popular on Semer Water, but not on Malham Tarn, which is kept as a nature reserve. The Wharfe appeals to many anglers as a very fine trout stream, while there are more than a dozen estates that offer grouse shooting, mainly on the grit moors. But two of the Dales great recreation activities are down to the geology – caving and climbing.

Ingleton is the busiest centre in Britain for cavers, simply because many of Britain's finest caves and potholes are crowded into the limestone plateaus around Gragareth, just to the north, and Ingleborough, just to the east. To many, a day's trip underground is a mixture of sport, adventure and excitement, but the regular Dales cavers soon get into searching for new cave passages. They have been dubbed 'weekend explorers'. They clear sediment chokes from ancient tunnels, they climb to unseen heights in the roofs of tall streamways, they squeeze into narrow rifts – all in order to discover new caves. In today's Britain, it is only in the limestone caves that anyone can be a genuine explorer, finding places never previously seen. But it is the efforts of these

Classic footpath erosion by trampling of boggy ground, and an engineered path that saves walkers and fell alike; both on the popular routes up Pen-y-ghent.

amateur cavers that have produced the maps of the caves and have yielded so much understanding of the Dales geology. A small slice of the underground adventure fraternity probes the old mines, especially around Swaledale; they too, through their recreation, find and record both the mining artefacts and the natural minerals that are components of the Dales underground.

Rock climbers cannot explore new ground within the Dales, but they do reach to some very inaccessible spots. Yorkshire limestone attracts large numbers of climbers to its vertical faces of clean and strong rock. Malham Cove, Gordale Scar and Kilnsey Crag are the Dales 'big three', each with over a hundred recognized routes, and Malham in particular is a veritable playground, which can attract dozens of acrobatic climbers to a warm summer afternoon in the sunlit amphitheatre of the Cove. There are also plenty of other lesser crags on the Great Scar Limestone, mainly where glaciers have trimmed a good strong bed. Twisleton Scars is one such, along the wall of Chapel-le-Dale's glaciated trough, though Yew Cogar Scar, above a tributary valley to Littondale, never had a powerful glacier scrape along its wall. Millstone Grit is famous for its rock climbing sites in the Pennines, but the best are all to the south of the Dales (and many of those are on crags steepened in bygone quarrying). Some parts of Brimham Rocks offer good sport, and Crookrise Crag is a popular site on the Grassington Grit of Embsay Moor. Elsewhere, certain isolated crags are left in peace as nesting sites for raptors, but cliffs of white limestone and dark grit are essential elements of the landscape throughout the Dales, appreciated from afar even if not examined from a tiny handhold.

Walking country

Some of the finest walking in England can be enjoyed within the Yorkshire Dales. The moors and fells offer senses of wilderness and dramatic isolation, yet the dissection by the dales creates ever-changing vistas; drama and detail are the essential ingredients of a good walk. Ramblers can find in the Dales a network of so many short footpaths, along with energetic routes like the Three Peaks (*see* page 215), and also sections of some popular long-distance trails that take spectacular routes through the Dales.

Footpaths within the Dales include a host of rights-of-way inherited from ancient usage. Mineral roads off the moors above Swaledale, and turbary roads down from peat bogs on many a fell, had geological origins among a host of packhorse trails, coaching routes, corpse ways and market roads. Few have evolved into metalled roads, while most have mellowed into countryside paths. They do include the Dales' famous 'green tracks'. Wider than roads, and penned between drystone walls, these are now highways for walkers. Mastiles Lane is the best known, from Malham eastwards to Wharfedale. This was once a drove road to Fountains Abbey, created for stock being brought in on the hoof from the outlying monastic farms; it has stayed green after plans to tar it were dropped when it was realized that it would become a mere rat-run on too many 'motor-tours'.

Footpath erosion is now an environmental threat. Across areas of wetter ground in the Dales, flagstone paths or raised boardwalks have become a necessity due to the huge numbers of walkers on some of the more popular routes. An unimproved trail across wet ground is soon trampled into a muddy bog, which deflects walkers around to trample and widen it yet further. And in the harsh Pennine climate, plant regeneration is very slow. Before the heavily used footpaths were engineered for durability, some had expanded into boggy swathes 50m (160ft) wide. These were most notable on the popular routes over the shale slopes high on Ingleborough and Pen-y-ghent. A few of today's paths may have lost their wilderness feeling, but the flagstones and boardwalks do allow people to enjoy the Dales countryside without destroying it. Paths on limestone are rarely a problem because the ground is so well drained, though that up to

Walkers head over snow towards Pen-y-ghent on the track from Littondale.

Boardwalk along the Pennine Way heading north toward the distinctive profile of Pen-y-ghent.

Malham Cove is a significant exception that required a good surface simply because of its popularity. When the steps up the side of the Cove were first built, they were declaimed as an eyesore, but they have now blended into the natural scars and look a lot better than would a swathe of muddy banks and degraded scree slopes.

Recent years have seen nature trails newly created for specific sites. Boardwalks around Malham Tarn give access to some very special wetland. A notable predecessor was the trail up Clapdale from Clapham village. The path now continues up onto the limestone fells and a variety of routes up Ingleborough, all of which are there through the hospitality of generations of the Farrer family who own the land. Apart from rights-of-way and a few commons, land on the high fells is all in private ownership. New legislation has designated as access land large swathes of unimproved country on the high limestone plateaus and open grit moors. As long as walkers, ramblers, cavers and climbers can respect the needs of the hill farmers, and as long as both can live in harmony, open access to so much the Dales can only increase the appreciation of these great landscapes.

Long-distance footpaths

Premier among England's footpaths, the Pennine Way traverses huge tracts of grit moors along the whole length of the Pennines, and gains some contrast when it crosses the best of the limestone country within the Dales. It emerges from the Craven Lowlands to head up Malham Dale and then take the steep route up the steps beside Malham Cove to enter the finest of the karst by tracing the floor of the Watlowes dry valley. After crossing the North Craven Fault, it winds round Malham Tarn and then rises through the limestone, and up onto the Yoredales to skirt the old coal mines on the summit of Fountains Fell. Over Pen-y-ghent, surely one of the finest viewpoints along the trail, and back down to the top of the Great Scar Limestone between two splendid potholes. Hunt Pot is a narrow rift with a stream dropping 50m (160ft) into darkness, while Hull Pot is a great open crater modified by collapse along its fault-guided walls and invaded by a large waterfall after heavy rain; both are worth short diversions. Along with all the other caves along the western side of both Pen-y-ghent and Fountains Fell, they both drain through to Brants Gill Head and also overflow to the flood resurgence nearby in Douk Gill.

North of Horton, the Pennine Way sees limestone scars vie with the finest of the drumlins, before a climb up and over Cam Fell. Follow the line of sinkholes in the Main Limestone and then a long descent to Hawes, through Yoredale rocks of course. Out past

Hardraw, for a view of the Force, the route climbs over the high grit ramparts of Great Shunner Fell, then crosses Swaledale with a traverse around the eastern flank of the Ice Age nunatek that is Kisdon Hill. Between Swinner Gill and Keld, tip heaps behind the old Crackpot Hall that mark the line of the Dales' greatest band of mineral veins, and then it's out onto the big grit moors. Old coal mines herald the Tan Hill Inn that fed miners long before the walkers came. Then a long, gentle, moorland descent, and across the natural span of God's Bridge to leave the Dales and head onwards to Scotland.

Not yet a designated long-distance footpath, the Coast-to-Coast Walk has become very popular along a fairly informal route across the grit country north of Swaledale. It enters

On its way north out of the Dales, the Pennine Way goes over God's Bridge, where the Yoredale Main Limestone provides a natural route over the River Greta that flows down from Stainmore.

The splendid view down Dentdale; on the skyline Great Coum is scored by its landslip scar that held a little corrie glacier just 12,000 years ago.

the Dales with a long climb from Kirkby Stephen across the Dent Fault and up most of the Carboniferous succession, to cross the Pennine watershed on the grit heights of Nine Standards Rigg. The remote Whitsun Dale takes the path down into upper Swaledale and past Wain Wath, Catrake and Kisdon Forces, all waterfalls on Yoredale rocks. Eastwards the route follows the main belt of mineral veins, marked by hushes, old tip heaps and ruined mills strewn across the moors either side of Gunnerside Gill. Down beside Mill Gill, the ruins of Old Gang Smelt Mill stand on wide tip heaps, while the Surrender Mill, further down the valley, has the better remains of its old flue up onto the high moor. The Coast-to-Coast then traces the floor of Swaledale between slopes and crags of the Main Limestone capped by cherts, before passing through Richmond and out of the Dales.

Third of the great trails is the Dales Way, which avoids the big hill climbs and takes an easier route along the floors of Wharfedale and Dentdale. Its Wharfe-side trail takes it past the grit ravine of the Strid, before it takes the high road north of Grassington, with a stroll through the splendid limestone pavements on Conistone Old Pasture. This offers views into the fine glaciated troughs of Wharfedale and Littondale, and the floor of the former is then followed north of Kettlewell. Into Langstrothdale, the valley profile is less rounded because it was scoured by less powerful ice. The limestone is left at Oughtershaw, and Yoredale slopes rise gently to the middle heights of Cam Fell, before a descent into the edge of the Ribblehead drumlin field. Newby Head is blanketed in drift except where a few dolines and caves trace the thin Yoredale limestones, before the descent into Dentdale passes under the black limestone arches of Dent Head Viaduct. Around Hackergill, the River Dee flows underground, and the Dales Way passes over the Tub Hole flood rising and the Popples main rising where it rejoins the river bank. The Dent Fault is crossed unseen beyond Dent village, and the Way soon leaves both Dentdale and the Dales as it cuts over a spur of Lake District slates to end at Sedbergh.

The Ribble Way and the Yoredale Way are two more long-distance footpaths with low-level profiles through the Dales. The Ribble Way traces up the wide floor of Ribblesdale, before cutting up through the best of the Ribblehead drumlin field, and ending where the headwater stream of Jam Sike emerges from the limestone cap above Cam Fell. As its name implies, the less well-known Yoredale Way follows the River Ure up the wide trough of Wensleydale to end on the watershed or continue down to Kirkby Stephen, after a traverse of the Dales practically all over Yoredale rocks.

Kirkby Fell, above Malham, is just one of the high fells that offer the very best of walking in wide-open spaces within the Yorkshire Dales National Park.

Walk the Three Peaks

With their steep profiles rising above the limestone plateaus between the southern Dales, the three peaks of Ingleborough, Pen-y-ghent and Whernside have become a magnet for keen hill-walkers. The classic Three Peaks Walk, with climbs of nearly 500m (1600ft) to the summits of each, is a circular route over 38km (24 miles) long, so it normally takes a full day of fairly hard walking. Or it can be run, in the annual fell race, when the top runners now make it round in well under three hours. Or each peak on its own can make a delightful half-day's walk, with time to enjoy the scenery.

Horton in Ribblesdale provides the traditional start, with a brisk climb straight to the top of Pen-y-ghent. From the long rise through the Great Scar Limestone, views directly behind are of the same beds eaten away by the huge Horton Quarry. Over a bit of a bench atop the limestone, the path continues onto a boardwalk over the wetter ground up slopes mainly of Yoredale shales. The path then joins the Pennine Way to climb the geological staircase up the southern shoulder of Pen-y-ghent. The first step is up the Main Limestone, while the second is up the Grassington Grit, and so to a gentle summit dome on mixed shales and grits. The views from the top are truly expansive – west across Ribblesdale to the limestone benches of Ingleborough, and east down the fluvial trench of Pen-y-ghent Gill into the glaciated trough of Littondale.

North off the summit, old workings for coal are barely recognizable, before the path drops steeply west down the Main Limestone, though much of its outcrop is covered by scree from the Grassington Grit. Down on gentler slopes the Three Peaks route cuts north away from the short route back to Horton, and crosses the beck upstream of its drop into Hull Pot. After avoiding the softest of the sphagnum bogs over Black Dubb Moss, drumlins dominate the landscape on the gentle descent into upper Ribblesdale. Beneath the first bench back on the Great Scar Limestone, Birkwith Cave has a fine walk-in exit just below the path, and Brew Gill is then crossed on another natural limestone span that is called God's Bridge. Gently down to cross a nascent River Ribble, and up to the road from

Winter walkers on the top of Pen-y-ghent look across Ribblesdale to the distinctive profile of Ingleborough.

Braithwaite Wife Hole lies in glacial till amid the limestone pavements of Southerscales Scars, seen from the path up Ingleborough, with Whernside in the distance.

where the best views of the Ribblehead drumlin field are off to the right.

Beside the towering viaduct, the path crosses Batty Moss where small shake-holes tell that this section of peat bog lies on a thin till that covers cavernous limestone beneath. It then follows the railway into Little Dale, past the old quarries that yielded dark Hardraw Scar Limestone to build the viaduct, then up Force Gill past waterfalls over each of the Yoredale limestones. A wide, peat-covered bench on the Main Limestone is overlooked by the summit ridge, from which blocks and slices of Grassington Grit have formed landslides down over the weaker shales between Grit and Limestone. Along the ridge above the landslide scars, Whernside's summit, highest point around the walk, offers a panorama across the Dales, and a view of the Lake District on a clear day.

South, off the summit, a ridge walk offers fine views south over the glaciated troughs of Chapel-le-Dale and Kingsdale with the lunar expanse of white limestone pavements on the intervening spur of Scales Moor. Descending from the ridge before the Combe Scar landslide, tiny sinks and springs down the steep hillside pick out the Yoredale bands of shales and limestones. The Great Scar Limestone is regained at Bruntscar, and the small river that drains Chapel-le-Dale is crossed unseen where it flows underground from the sinks at Haws Gill Wheel to the rising at yet another God's Bridge.

Leaving the road just above the Hill Inn, the best path up Ingleborough heads off across delightful springy turf along a limestone bench, and offers fine views down Chapel-le-Dale. A perfect U-shape profile of the glaciated trough is seen almost in silhouette where it breaks out from the limestone across the distant Craven Faults. On each side, and rising gently with the dip, the main benches are bare limestone pavements where they were scraped clean by the Ice Age glacier. The path eventually turns up through the Southerscales Scars, where the best of the splendid limestone pavements are only seen by looping, left or right, away from the shortest line. Braithwaite Wife Hole, a large and ancient doline half-filled with glacial till, is passed before crossing a wall almost at the edge of the limestone, before a long climb up the Yoredale slopes. The saddle between Ingleborough and Simon Fell is reached just below the line of white scars on the Main Limestone, and the path then rises over these and up onto the Grassington Grit.

Ingleborough's summit is an almost triangular plateau on just the lower bed of the Grit. A bleak and windswept site indeed, it was ideal for the Iron Age hill-fort last occupied by the Brigantes tribe who were resisting Roman colonialism nearly 2000 years ago, for it commands fine views in all directions. Wide limestone pavements dominate to the northwest, broken by the deep glaciated troughs of Chapel-le-Dale and Kingsdale. South looks across the Craven Faults and out over the Craven Lowlands, to the rolling hills of the Bowland Forest. To the east, the distinctive profile of Pen-y-ghent is a landmark in front of the endless hills and ridges that rise between the finest of the Dales.

Back down to the Simon Fell saddle, the last leg of the route passes the springs from the Main Limestone that kept the hill fort supplied. Then on down the Yoredale slopes to meet the Great Scar Limestone again where Sulber Pot has the only large entrance among various deep potholes. From there, Sulber Nick lies along the line of a small fault, and its grassy trench provides the easiest line through the limestone pavements. Descend through the scars to the floor of Ribblesdale and pass a few more drumlins on the final return to Horton village. Whether the day's walk has taken in one peak or all three, it has traversed some of the best of the geology, the karst and the glacial landforms that combine so well in the grand scheme of the Dales landscape.

Wide limestone pavements form much of the benches on top of the Great Scar Limestone on the east side of Ingleborough, while in the distance Pen-y-ghent rises through the Yoredales to its Millstone Grit cap.

Further Reading

Information may be derived from a host of sources, among which are a few key publications. A mass of geological data lies within *Geology of the Northern Pennine Orefield* (volume 2, *Stainmore to Craven*, by K. C. Dunham and A. A. Wilson, 1985, British Geological Survey, HMSO, London, 247 pages). Details on most of the karst landscapes and caves are in *Karst and Caves of Great Britain* (Geological Conservation Review Series 12, by A. C. Waltham, M. J. Simms, A. R. Farrant and H. S. Goldie, 1997, Chapman & Hall, London, 358 pages). There is no comprehensive published summary of the glacial development of the Dales landscapes; information is spread through numerous books and academic papers. The classic text on people in the Dales is *A Dales Heritage* (by Marie Hartley and Joan Ingleby, 1982, Dalesman, Clapham, 142 pages). The National Park website at yorkshiredales.org.uk has some excellent material that is both authoritative and accessible.

Glossary

Alluvium Sediments and soils left by a river along a valley floor.

Anhydrite Calcium sulphate deposited as a mineral in some shallow lagoons.

Anticline Upward fold in bedded rocks, caused by crumpling during earth movements.

Basement The older rocks, commonly metamorphosed, beneath a cover of younger sedimentary rocks.

Brachiopod Bivalve sea animal with two thick shells, which is common as fossils in limestone.

Breccia Sedimentary rock with coarse angular fragments, like scree debris.

Carbonate Can be used as a general term to include all the limestone and dolomite rocks that are made of calcium and magnesium carbonates.

Carboniferous Geological time from 359 to 299 million years ago.

Chert Very hard, fine-grained rock made of pure silica.

Clastic Descriptive of rocks made of derived particles of minerals, therefore including sands, sandstones, muds and clays.

Clint Block of limestone bordered by open fissures (grykes) within a limestone pavement.

Colluvium Collective term for all types of soils and broken rock on slopes.

Conglomerate Sedimentary rock with coarse rounded fragments, like a pebble bed.

Connate water Water trapped in sediments and then squeezed out as the sediment is compacted into sedimentary rock.

Corrie A hillside bowl that was once occupied by a small glacier; also known as a cirque or cwm.

Crinoid Animal related to the sea urchin, but standing on one long leg so that it looks like a plant; common as fossils in limestone.

Cyclothem A cycle or repeated succession of a few sedimentary rocks, originally formed when a cycle of earth movements caused the changes in sedimentation.

Debris flow A slow-moving avalanche of rock debris down a slope.

Devonian Geological time from 416 to 359 million years ago.

Doline A closed depression in the ground surface, where water sinks underground, and therefore typical of limestone terrains.

Drift Collective term for all unconsolidated sediments and soils that overly solid rock.

Dripstone Collective term for cave deposits formed by dripping water, therefore including stalactites and stalagmites.

Drumlin A rounded hill of glacial debris (*see* the box on pages 84–85).

Erratic An isolated large boulder that was carried to its present position by an Ice Age glacier.

Evaporite Collective term for minerals and rocks deposited by partial evaporation of water, therefore including salt and anhydrite.

Fault A break in the rock structure where one side has slipped past the other (and caused an earthquake when the movement was a sudden jolt).

Flowstone Collective term for cave deposits formed by flowing water, therefore including floor deposits and some types of stalagmite.

Gangue The uneconomic minerals within a vein that is being mined for its economically valuable ore minerals.

Glacial striae Parallel scratches on a rock surface created by boulders being dragged along in the basal ice of an Ice Age glacier.

Graben Block of rock, typically kilometres across, that has dropped down between two major faults.

Greywacke Strong, old slightly metamorphosed sandstone, commonly found interbedded with slate.

Gryke Fissure, enlarged by rock dissolution from an initial fracture, that lies between blocks of limestone (clints) within a limestone pavement.

Horizon Bedding plane that is a time boundary within a sedimentary rock sequence, though it may not now be horizontal.

Horst Block of rock, typically kilometres across, that has been uplifted between two major faults.

Hush A channel washed out of the soil by miners searching for or extracting mineral veins (*see* the box on page 165).

Ice Age Period of time when worldwide cooling allowed glaciers to cover many areas, including the Yorkshire Dales; a stage within the Pleistocene.

Inlier Outcrop of old rock surrounded by younger rocks, commonly found along a valley floor.

Joint A fracture in rock, created by natural earth movements, along which there has not been movement (so it is not a fault).

Kame Small hill of sediment deposited by a meltwater stream at the base of an Ice Age glacier.

Karren Collective term for the long and narrow grooves carved into limestone by trickling rainwater.

Karst A landscape distinguished by underground drainage, notably though caves, and therefore typical of limestone terrains.

Loess Wind-blown silt, mostly blown from areas adjacent to Ice Age glaciers when there was no plant cover to bind the soil.

Mesozoic Geological time from 251 to 65 million years ago.

Moraine A layer, hill or ridge of till.

Ordovician Geological time from 488 to 443 million years ago.

Ore Mineral within a vein or rock that can be mined economically.

Orogeny A phase of earth movements when folding and faulting took place related to the collision of two continental plates.

Palaeozoic Geological time from 540 to 251 million years ago.

Periglacial Peripheral to a glacier, therefore descriptive of terrain with frozen ground but not covered by the glaciers.

Permian Geological time from 299 to 251 million years ago.

Phreatic Beneath the water table, where rock voids are full of water; including caves that have developed beneath the water table.

Pleistocene Geological time that includes the Ice Ages, between 1,800,000 and 11,500 years ago.

Precambrian Geological time before 540 million years ago.

Quaternary Geological time from 1,800,000 years ago to the present day, therefore including the Pleistocene, the Ice Ages and post-glacial times.

Reef knoll Hill or knoll of strong limestone that largely follows the shape of an original reef (composed of corals or algae).

Rugose coral A type of coral that forms an individual cup or cylinder as opposed to the colonial corals.

Sandstone Sedimentary rock that is a mass of sand grains held together by a natural mineral cement.

Scree or **talus** Synonyms for the angular rock debris that forms aprons below cliffs where frost-shattered bits of rock have fallen from above.

Shale Sedimentary rock formed by a mass of clay particles that bond together when the water has been squeezed out from between them.

Silurian Geological time from 443 to 416 million years ago.

Solifluction Downslope movement of saturated soil, a type of rapid soil creep common on slopes in periglacial environments.

Stope Open cavity created by miners removing the minerals from a vein.

Stratigraphy Sequence of sedimentary rocks that record a history of deposition.

Stratimorph Section of the ground surface that lies on a single bed (or stratum) of resistant rock.

Syncline Downward fold in bedded rocks, caused by crumpling during earth movements.

Terracette Small ledge or terrace in the soil of a steep hillside, formed by soil creep and enhanced by walking sheep.

Tertiary Geological time from 65 to 1.8 million years ago.

Till Unsorted debris left by a glacier; also known as boulder clay.

Tor Isolated hill or crag of rock that is less jointed than the surrounding material, which has since been weathered away.

Triassic Geological time from 251 to 199 million years ago.

Tufa Calcium carbonate deposited by algal action in a stream bed in limestone country; similar to stalagmite but not in a cave.

Unconformity Top surface of old folded rocks that were partly eroded away before younger sedimentary rocks were deposited over them, and therefore indicative of a big time gap in the stratigraphical record.

Vadose Above the water table, where drainage flows freely downwards through a rock mass; including caves that have developed above the water table.

Index

Entries in **bold** denote illustrations